On-Chip Electro-Static Discharge (ESD) Protection For Radio-Frequency Integrated Circuits

Qiang Cui • Juin J. Liou • Jean-Jacques Hajjar
Javier Salcedo • Yuanzhong Zhou
Srivatsan Parthasarathy

On-Chip Electro-Static Discharge (ESD) Protection For Radio-Frequency Integrated Circuits

Springer

Qiang Cui
Qorvo Inc.
Greensboro
North Carolina
USA

Juin J. Liou
University of Central Florida
Orlando
Florida
USA

Jean-Jacques Hajjar
Analog Devices Inc.
Wilmington
Massachusetts
USA

Javier Salcedo
Analog Devices Inc.
Wilmington
Massachusetts
USA

Yuanzhong Zhou
Analog Devices Inc.
Wilmington
Massachusetts
USA

Srivatsan Parthasarathy
Analog Devices Inc.
Wilmington
Massachusetts
USA

ISBN 978-3-319-10818-6 ISBN 978-3-319-10819-3 (eBook)
DOI 10.1007/978-3-319-10819-3

Library of Congress Control Number: 2015931918

Springer Cham Heidelberg New York Dordrecht London

Printed on acid-free paper

Springer is a brand of Springer International Publishing
Springer International Publishing is part of Springer Science+Business Media (www.springer.com)

Love is life's wonder! I dedicate this book to my wife Yan Zhou.

By Qiang Cui

Introduction

Electrostatic discharge (ESD) is a pervasive physical phenomenon with the potential of irreversibly damaging integrated circuit throughout the manufacturing cycle, from semiconductor wafer processing through dicing and package assembly as well as through integration into the end product. Several ESD stress models and test methods have been used to reproduce ESD events and characterize ESD protection device's performance. The most common ESD stress models are the human body model (HBM) and the charged device model (CDM). On-chip protection devices as well as circuitry are commonly used to shunt ESD current and limit overstress. To be effective, ESD protection devices have to be transparent to the performance of the circuit and trigger only during an unintentional electrostatic discharge event. A wide variety of such devices have been developed for different process technologies from complementary metal–oxide–semiconductor (CMOS) to complementary bipolar to bipolar CMOS (BiCMOS) to III-V compounds.

Achieving ESD robustness on high-performance circuits has been a challenge. This is particularly true for radio frequency integrated circuits (RF ICs) where the addition of any protection device to an RF circuit node will interfere with the functionality of the circuit to be protected. ESD protection for RF ICs is more demanding than traditional low-speed CMOS ESD protection design due to IC manufacturing process and intrinsic parasitic capacitance limitations. High-performance RF ICs are partial to compound semiconductor manufacturing processes, such as gallium arsenide pseudomorphic high-electron mobility transistor (GaAs pHEMT) or silicon–germanium heterojunction bipolar transistor (SiGe HBT). Proven and robust ESD devices (e.g., silicon controlled rectifier, SCR) cannot be fabricated in the latter processes due to the absence of the building blocks that constitute the device in question. Additionally compound semiconductor processes have lower thermal conductivity making them more susceptible to ESD damage. On the other hand, the intrinsic parasitic capacitance of ESD protection devices is a concern requiring a delicate balance between robustness and circuit performance. This book's focus will therefore be on ESD protection designs for RF ICs.

Chapter 1 introduces the fundamentals of ESD with a brief review of the many different ESD testing methodologies. Basic ESD protection design and circuit architecture methodologies are introduced. The challenges of protecting RF circuit are also addressed. Chapter 2, continues the discussion on RF ESD protection design with a focus on standard baseline CMOS process technologies. The importance of low-capacitance protection structures is discussed and such devices, in the form of silicon controlled rectifiers among others, are introduced. The methodology of reducing the parasitic capacitance of these devices and a discussion of the underlying physical principles are reviewed. Chapter 3 surveys the ESD protection approach on SiGe–BiCMOS processes: a class of process technology used extensively to build RF circuits. The focus of this chapter is on SCRs (SiGe SCR). SiGe SCRs are good candidates for ESD protection device on these processes as they provide good robustness with high current shunting capability per unit area and consequently occupy a small footprint in the circuit layout. This is highly desirable as it reduces the intrinsic parasitic capacitance of the device. Nonetheless, these devices may be slow to turn on, which under ESD CDM stress conditions may not be very efficient. The chapter demonstrates how technology computer-aided design (TCAD) simulations along with characterization under very fast transient pulse stresses (i.e., <5 ns) are leveraged to achieve the desired device performance. Engineering the device to reduce voltage overshoot during the initial ESD stress is reviewed. This serves as a practical approach for designing optimum ESD protection solutions for the low-voltage/radio frequency integrated circuits in SiGe BiCMOS process.

Chapter 4 is dedicated to RF ESD protection in III-V compound semiconductor process technologies and in particular GaAs pHEMT. The latter processes offer very little in terms of building blocks to design proper RF ESD protection devices. The chapter analyzes pHEMT's snapback, postsnapback saturation, and thermal failure under ESD stress using TCAD simulations. The snapback is caused by a virtual bipolar transistor that is formed due to a large electron–hole pair concentration near the drain region. Postsnapback saturation, on the other hand, is the result of temperature-induced mobility degradation due to III-V compound semiconductor materials' poor thermal conductivity and hot spots located in pHEMT's indium gallium arsenide (InGaAs) layer. This induces thermal failure in the device. Understanding these physical mechanisms is shown to be critical in designing an effective ESD protection device in GaAs pHEMT processes. Several novel ESD protection devices designed on 0.5-µm GaAs pHEMT process are introduced for illustration purposes. A multigate pHEMT based ESD protection device in both enhancement mode and depletion mode is introduced. Due to the multiple current paths available in the multigate pHEMT, the new ESD protection clamp shows significantly improved ESD performances over the conventional single-gate pHEMT ESD clamp, including higher current discharge capability, lower on-state resistance, and smaller voltage transient. Further enhanced ESD protection clamp based on a novel drainless, multigate pHEMT in a 0.5-µm GaAs pHEMT technology is also presented. It is shown that through proper design the HBM classification test levels of the ESD protection levels of the may be tuned from 1000 V to over 8000 V.

Finally, Chap. 5 summarizes the fundamental principles related to ESD protection for RF circuits. It also provides a unified guideline on how to achieve such robustness across process technologies and the tools required to support such an objective. Insights and direction into future work in this area are also discussed.

Contents

List of Figures

List of Tables

Chapter 1
Basics in ESD Protection of Radio Frequency Integrated Circuits

Electrostatic discharge (ESD) has been a critical design concern to integrated circuits (ICs) for many years [1–10]. ESD has posed a major product reliability threat as it is present in all IC manufacturing processes from mainstream silicon-based complementary metal–oxide–semiconductor (CMOS) to the more exotic compound semiconductor processes such as gallium arsenide (GaAs) or gallium nitride (GaN).

Static electricity has been prevalent for ages. It was first recorded over a couple of thousand years ago when people observed that amber when rubbed against one's skin was able to pick up light objects such as feathers. The phenomenon of electrical charge was not understood, however, until the eighteenth century by the likes of Coulomb as well as Gauss, and of course by Benjamin Franklin with his famous experiment where he flew a kite in a storm to harness electricity from a thunderstorm.

Static electricity is the build up of charge on the surface of object. These charges remain on the object, and may be transferred to other objects. The charge eventually bleeds off to a grounded surface or is quickly neutralized by a discharge.

The physical phenomenon that describes this static charge buildup and exchange is called the triboelectric effect. Two neutral materials when brought into physical contact with each other and then separated will exchange electron, leaving one material negatively charged and the other positively charged. The exchange of electrons depends on many factors such as room temperature, relative humidity, and duration of contact, among others. The key parameter, however, is the materials in question. A good example is the charge generated when rubber balloon is rubbed on one's hair. The hair will have a tendency to shed electrons to the balloon.

In a semiconductor fabrication environment, the said charge could be induced by human operators, the manufacturing machine/equipment, or by the semiconductor device itself. The transfer of static charge from one surface to another happens very quickly and results in large transient voltages and currents. The latter may be very harmful to the IC as it could cause damage to the MOS transistor gate oxide, the on-chip metallization, as well as the p–n junctions. Steve Halperin, an ESD consultant and past president of the ESD Association, conducted a study a few years back where he showed that poor ESD robustness was costing the industry as much as 8 % of total revenues in lost profits [11]. The author catalogued the effect of poor ESD robustness as follows:

Q. Cui et al., *On-Chip Electro-Static Discharge (ESD) Protection For Radio-Frequency Integrated Circuits*, DOI 10.1007/978-3-319-10819-3_1

i. Redesigns, which delays product release
ii. Unsatisfied customers, which implies business hold
iii. Competitive disadvantage, which results in loss of business
iv. High test yield loss, which cuts into profit
v. High customer return rates, which results in more product analysis support

Traditionally, there have been two approaches to mitigate the effects of ESD; and the two are not mutually exclusive, in fact they work in tandem. The first approach is to control ESD in the manufacturing environment. Proper gears, such as grounded strap or special gloves by human operators, ionizers to neutralize static electricity, and antistatic material for the floor on the production facility, have been extensively adopted. The second approach is to protect the IC circuits from electrostatic discharge by making them more robust. This is achieved through the design of special protection circuitry on-chip.

ESD protection devices and circuits for low speed/low frequency circuits are fairly mature and the typical ESD failure level is above 2000 V human body model (HBM) classification test level. However, effective ESD protection for radio frequency is still not widely prevalent [9, 12, 13].

i. Existing ESD protection solutions have been developed mostly for CMOS processes. There are basically no such solutions for compound semiconductor processes such as GaAs pHEMT and SiGe HBT. Compound semiconductor processes are widely used in performance RF/microwave circuits, and they are generally vulnerable to ESD events because of the inherent low thermal conductivity of the compound semiconductor materials. The state of the art in ESD protection technology for these processes is only 1000 V ESD HBM classification test level and is achieved through the use of stacked Schottky diode-based circuit architecture.
ii. Even though ESD protection solutions for RF ICs in CMOS process have been demonstrated, the said solutions suffer from relatively large parasitic capacitances. This makes them suitable for only a handful of applications.

There is a growing demand for the availability of robust ESD protection solutions for radio frequency integrated circuits (RFICs). This book will address device design and device simulation of ESD protection devices in compound semiconductor processes (GaAs pHEMT and SiGe HBT). Examples of ESD protection devices in CMOS, BiCMOS, as well as compound semiconductor process technologies will be discussed in the subsequent chapters of this book.

1.1 ESD Models and Test Setup

ESD events have been classified into three different categories, each representing a certain physical phenomenon. The models developed for this purpose are: HBM, machine model (MM), and charged device model (CDM). Figure 1.1a, b and c depicts the discharge waveforms associated with each of the models mentioned above.

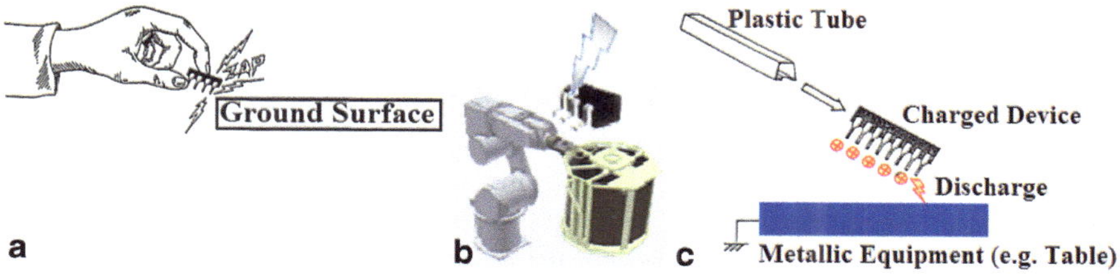

Fig. 1.1 Discharge model for: (**a**) human body model (HBM); (**b**) machine model (MM); and (**c**) charged device model (CDM)

HBM is the most commonly used model during reliability testing. It is a requirement for any IC component. The model mimics the discharge from a human operator handling ICs. The charge accumulated on a human, as a result, for instance, of walking across a carpeted floor on a dry winter day, discharges through any object that the said person touches with his finger. The human in this case would be the source of charge and is represented by a capacitor and the discharge that would presumably be through the finger of the human is represented by a series resistance of 1500 Ω. Obviously these parameters would vary depending on many conditions such as the humidity of the environment and the constitution of the human. Nonetheless, a standard was drafted for this model and is documented in MIL-STD-883; "Test Method Standard Microcircuits" [14]. One of the key characteristics of the HBM stress is that it lasts only a few hundreds of nanoseconds, it is unipolar in nature (i.e., either positive or negative), and its peak current could reach several amperes.

MM simulates the discharge from a charged conductive source to an IC. Common examples of this model would be the discharges from in-line IC inspection equipment, not properly configured for ESD control methods, or nongrounded areas of automatic test equipment (ATE). There are several independent standards that describe the MM. They include the JEDEC Solid State Technology Association JESD22-A115 [15], the ESD Association ANSI/ESD-STM5.2 [16], the International Electrotechnical Commission IEC-60749-27 [17], and the Automotive Electronics Council's AEC-Q100 [18], among others.

The stress model that has been proposed for the MM is similar to all other classification test models. In its basic form, it consists of a lumped RLC (resistor, inductor, and capacitor) circuit, the circuit schematic of which is shown in Fig. 1.1b. This is similar to the HBM with the exception of the circuit element values. In the case of the MM, since the source is conductive, the RESD value is, nominally, 0 Ω. This circuit configuration results in a waveform with large amplitude, submicrosecond duration, and low damped decaying sinusoid. In practice, however, this waveform is impossible to achieve, since the additional parasitic in the tester system that connects to the IC will come into play. The parasitic has a strong influence on the discharge current waveform, especially since the IC will have a dynamically changing impedance as its protection turns on in response to the ESD pulse. As a result, the waveform variation from pin-to-pin within an IC will be appreciable. Industry support for the MM has been in decline for years. This is attributed to many factors. Among them is the poor test-level correlation that has been reported between various

MM ESD test systems, the lack of a unique MM-specific failure signature, and the high cost of qualifying an IC through different ESD classification tests. The MM has come to be viewed as simply a more severe HBM stress. Moreover, its failure signature has a very high degree of correlation (well over 90 %) to that of the HBM stress. In light of this, The ANSI/ESD S5.2 standard document is being downgraded to a standard test method (STM). As an STM, the MM document is no longer allowed for use in qualification of devices using MM testing.

CDM describes charging/discharging events that occur within the IC components and semiconductor itself. One of many examples of such a phenomenon is a device sliding down a shipping tube and coming into contact with a metal grounded surface. The discharge current is limited only by the parasitic impedance and capacitance of the device. Therefore, the discharge current pulse is very fast (typically a few nanoseconds) and is several amperes in magnitude (i.e., over 10 A is very common). According to a survey, over 99 % of ESD damage in ICs is caused by the CDM stress mechanism [19].

The equivalent circuit representation of various ESD event models is depicted in Fig. 1.2a [20]. All the three models (HBM, MM, and CDM) may be described by one equivalent circuit topography (see Fig. 1.2) with the values of the elements being different. The current waveforms of HBM, MM, and CDM result from different RLC values. If the oscillation frequency $\omega = 1\sqrt{(LC)}$ exceeds the damping coefficient $\alpha = R/2L$, including the load resistance, the discharge is an oscillation as observed for the MM and CDM with low resistive loads. Otherwise, it is periodically damped, like the HBM.

All the above ESD classification test models are used to test the device through to failure. Figure 1.3a depicts a transient HBM tester that tests to failure at the wafer level and Fig. 1.3b is an ESD "gun" used for a similar purpose except at a system level instead of the component level. However, these destructive test methods cannot supply insight into the device failure mechanism. To overcome this disadvantage, a testing method called transmission line pulse (TLP) was first introduced by Tim Maloney to mimic HBM ESD events [21]. Figure 1.4a describes the principle of TLP measurements. A transmission line cable is used to generate the stable square current waveforms that discharge into an IC component. This simulates the ESD stress. The TLP pulse is typically 100 ns in duration and has a varying pulse rise time that ranges from 1 to 10 ns. More recently, the above test method has also

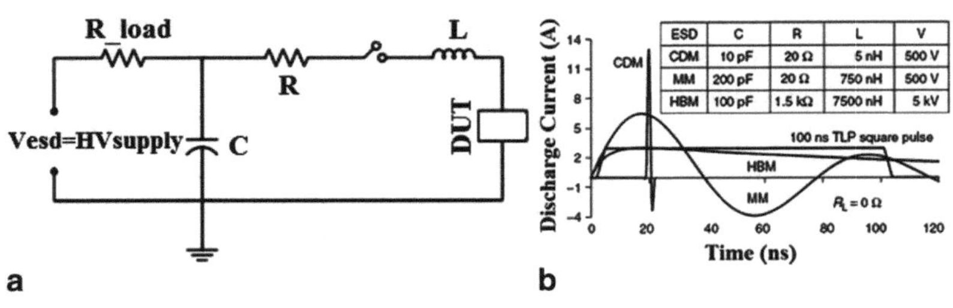

Fig. 1.2 **a** RLC equivalent circuit for ESD classification test models. **b** Discharge current waveform of ESD classification test models in Fig. 1.1

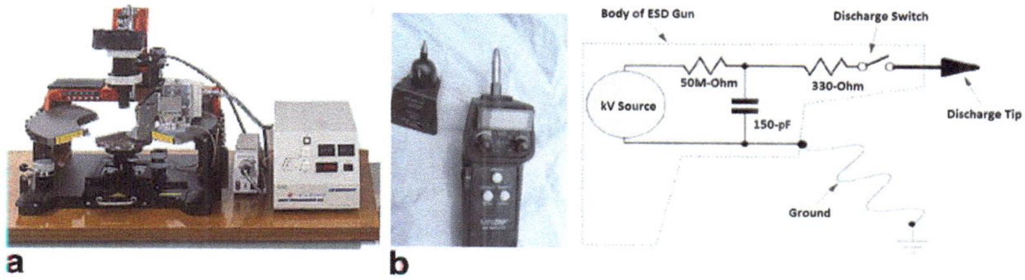

Fig. 1.3 **a** Wafer-level transient HBM tester from Hanwa (model HED W5000M). **b** System-level ESD-gun from Schaffner/Teseq

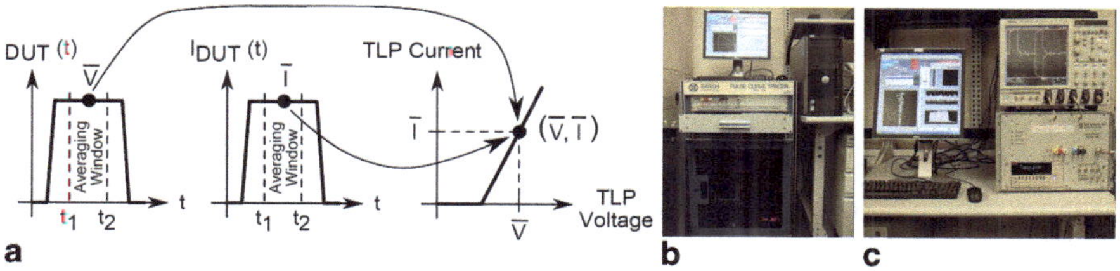

Fig. 1.4 **a** Transmission line pulse (TLP) measurement method for ESD protection devices characterization. **b** Barth 4002 TLP measurement system for HBM ESD classification test model characterization. **c** Barth 4012 TLP measurement system for CDM ESD classification test model characterization

been extended to much shorter pulses. The very fast TLP (vfTLP) system was developed to mimic CDM events. The vfTLP test system is similar in concept to the standard TLP system but it has a much narrower pulse width (1–5 ns) and shorter pulse rise time (0.1–0.5 ns).

1.2 On-Chip ESD Design Concepts and Design Windows

In order to protect internal circuits against unintentional ESD damage, ESD protection networks are designed and embedded on-chip in the circuit. A generic ESD protection topography is shown in Fig. 1.5 [22]. This protection network consists of five essential components. They are:

1. Input pad to ground "Vss" protection
2. Input pad to power rail "Vdd" protection
3. Output pad to ground "Vss" protection
4. Output pad to power rail "Vdd" protection
5. Power rail "Vdd" to ground "Vss"

The protection device/circuitry should be transparent to normal circuit operation. The protection circuitry should "turn on" only during an ESD event. This on-chip

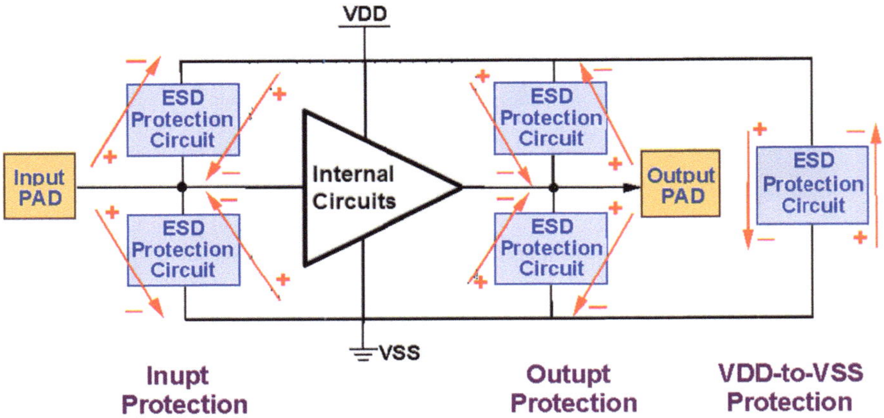

Fig. 1.5 Generic on-chip ESD protection network

ESD protection network is a generic strategy employed on all major processes including CMOS, BiCMOS, as well as III-V compound.

1.2.1 Characteristics of ESD Protection Devices

On-chip ESD protection devices rely on their physical size to sink the large discharge current, albeit for a very short period, and maintain the resulting voltage drop at a reasonable level. In addition, they also rely on the device's physical property during electrical breakdown. This is generally referred to as "snapback" [23]. In the latter case, the electrical breakdown generates a low-conductivity path, allowing large current discharges without incurring high voltages. Not all the ESD protection devices have "snapback" characteristics. The electrical characteristics of the two types of devices are shown in Figs. 1.6 and 1.7, respectively.

In the first instance, the ESD devices are characterized by a threshold point (Vt1, It1). When the applied voltage across the device is below Vt1, the device is in a very high impedance state and negligible current flows through it. Beyond the threshold voltage, Vt1, a low-impedance path is created, which facilitates the discharge of the ESD current. As the current increases further, it will reach a point where the device will sustain an unrecoverable thermal damage. This is described by data point (Vt2, It2) in Fig. 1.6 and is usually referred to as the secondary breakdown. The ESD design window for this type of ESD devices is shown in Fig. 1.6b Different colors are used to differentiate three regions: the internal circuit's operation region (blue), the internal circuit's failure region (red), and the ESD device's failure region (yellow). The ESD design window is surrounded by three regions and highlighted by the solid green lines. A good ESD turn-on characteristics protection device should have its I–V characteristics within this ESD design window. P–N diodes in CMOS processes and Schottky diodes in GaAs pHEMT process are only two such examples of devices used for ESD protection.

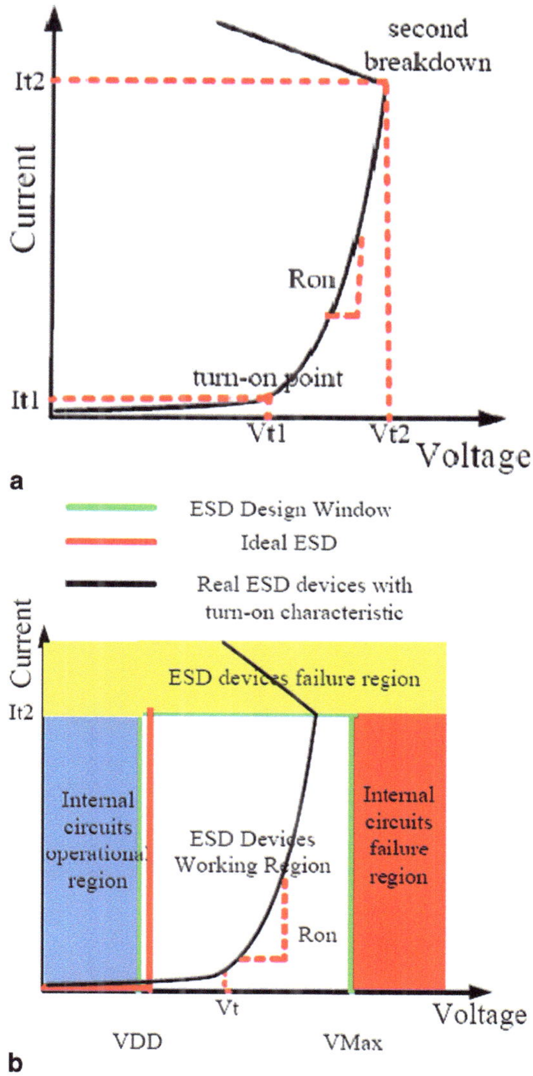

Fig. 1.6 ESD protection devices with turn-on characteristics. **a** Typical I–V curve. **b** ESD design window

Figure 1.7a shows the I–V characteristics of ESD devices that exhibit "snapback" characteristics. This type of ESD devices has similar characteristics to the ones described above before reaching the trigger voltage (Vt1, It1). However, when the voltage across the device exceeds this trigger voltage Vt1, its I–V characteristics will be forced into "snapback." The voltage across the device drops dramatically and a low resistivity discharge path is created. Similarly, with increasing current and voltage, the device reaches its secondary breakdown point where it will experience thermal failure. The similar ESD design window for this type of ESD devices is

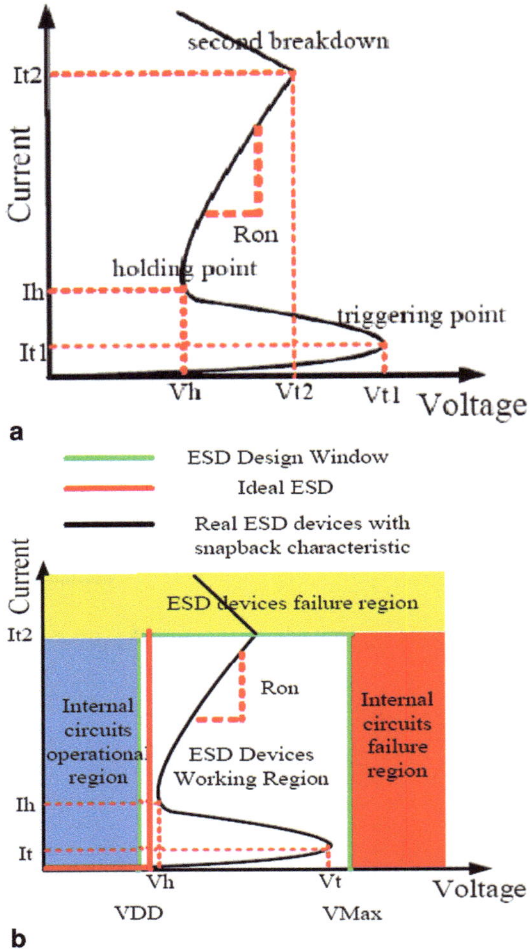

Fig. 1.7 ESD protection devices with snapback characteristics. **a** Typical I–V curve. **b** ESD design window

shown in Fig. 1.7b. A good ESD snapback characteristics protection device should have its I–V characteristics located in this ESD design window. Examples of snapback characteristics ESD device include externally triggered active clamp in GaAs pHEMT process and silicon controlled rectifier (SCR) in CMOS process.

1.2.2 Typical On-Chip ESD Protection Devices for Radio Frequency Integrated Circuits

A diode chain is simple and widely used as ESD protection device for RF ICs in both CMOS and GaAs pHEMT processes as shown in Fig. 1.8. The diode chain could be used to protect low voltage I/O pins, or combined with power supply

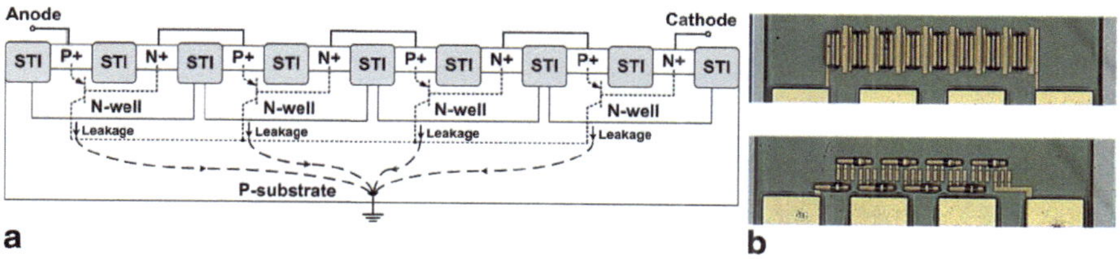

Fig. 1.8 **a** Conventional diode chain in CMOS process. **b** Schottky diode chain in GaAs pHEMT process

clamps to provide rail-based ESD protection. The main advantage of diode chains is that they have fairly low parasitic capacitance. Moreover, by connecting the diodes in series as part of an ESD diode protection chain, the parasitic capacitance may be decreased further. This, however, comes at the expense of additional footprint of protection device. Moreover, the diode exhibits poor leakage current characteristics in its "OFF" state. There are many reports in the literature that attempt to optimize the diode for on-state resistance while mitigating the leakage current [24]. Additional devices are used in order to block the leakage current path.

Externally triggered ESD clamp as shown in Fig. 1.9 is an alternative protection circuit for RF ICs in GaAs pHEMT process. Because of its relatively higher ESD current handling ability, the externally triggered ESD clamp is a more favorable protection device in GaAs pHEMT process. The trigger voltage of ESD clamp could be adjusted by changing the number and orientation of the trigger diodes. Such clamps have been reported in the literature with robustness exceeding 1000 V equivalent HBM classification test levels [25].

SCR in Fig. 1.10 is another type of ESD protection device. It consists of successive p-n-p-n diffusions or two interconnected NPN and PNP bipolar transistors. When the SCR turns on, both PNP and NPN transistors conduct as a result of carrier injection into the bases of both devices. The corresponding conductivity modulation

Fig. 1.9 Externally triggered ESD clamp in GaAs pHEMT process

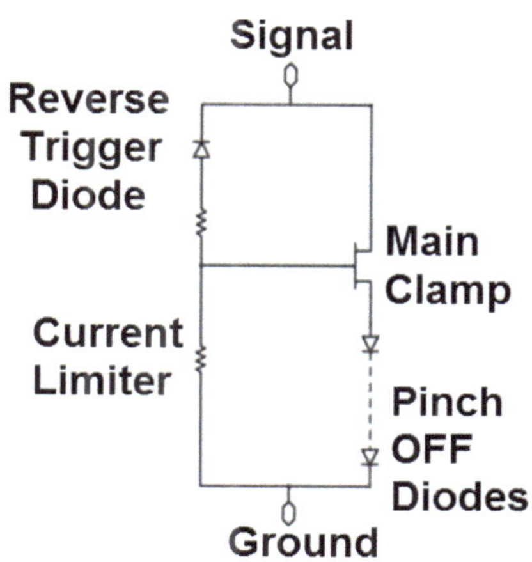

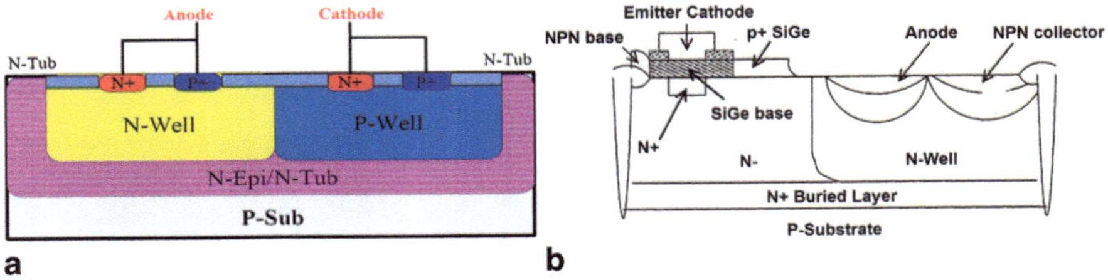

Fig. 1.10 Cross section of: **a** Lateral SCR in CMOS process; **b** vertical SCR in SiGe BiCMOS process

provides a low impedance path for the current to discharge. In light of this, the SCR has a much higher current handling ability per unit area than any other device. This makes for an excellent ESD protection device. However, there are several disadvantages of using SCR as ESD protection devices:

1. The SCR is not possible to be fabricated in GaAs pHEMT process because only N-type is available in such a process.
2. The SCR requires a large trigger voltage.
3. The SCR voltage collapses to a very low value during "snapback," which may not be desirable and may also constitute a latch-up risk.
4. The SCR exhibits a slow turn-on speed with large overshoot voltage.

Both lateral SCR (see Fig. 1.10a) in CMOS process and vertical SCR (see Fig. 1.10b) in SiGe BiCMOS process are described in Chap. 3. Vertical SCRs are considered to have faster turn-on speed due to the high efficiency (high-Beta) of vertical NPN heterojunction transistor, and it will be further optimized in Chap. 3 in SiGe BiCMOS process.

1.3 Special Challenge of ESD Protection in Radio Frequency Integrated Circuits

The fundamental challenge of ESD protection design for RF ICs is the requirement of low parasitic capacitance. ESD protection unit is normally put at input/output of ICs, the parasitic capacitance could distort the RF performance such as noise figure, linearity, and power handling ability. There are other special considerations for RF ESD protection design including constant capacitance over frequency and mistrigger immunity against the substrate noise. The key to a successful RF ESD protection architecture is a codesign. The protection device/cell should be accounted in the design of the RF pin. This requires a careful characterization of the said device that goes beyond the simple ESD characterization methodologies described above. The high frequency electrical characteristics of the ESD device are critical in this respect.

The next chapter will address in more details the concepts reported above.

1.4 Summary

As the feature size and, consequently, the complexity of semiconductor fabrication technology continue to increase, ESD-induced reliability issues are becoming more challenging to resolve. In order to reduce IC failure caused by ESD damage, on-chip ESD protection devices are widely used to discharge ESD current and limit the overstress voltage under different ESD events. There are only a few ESD protection devices available for RF ICs, and RF ESD design is more challenging than traditional low speed CMOS ESD protection design. This chapter introduced ESD phenomena fundamentals, ESD models, ESD test methods, and different types of ESD device's characteristics. And also, we introduced three on-chip ESD protection devices available in existing literature: (1) diode and diode chain, (2) externally triggered ESD clamp, and (3) SCR.

Chapter 2
On-Chip Protection Solution for Radio Frequency Integrated Circuits in Standard CMOS Process

2.1 Introduction

Standard CMOS technologies have been increasingly used in RF IC applications mainly due to low manufacture cost and improvement in performance. Achieving sufficient ESD protection for RF and high-speed mixed signal ICs using mainstream CMOS processes imposes a major design and reliability challenge. Ideally ESD protection must be transparent to the protected core circuitry under normal operation conditions. In reality, interaction always exists between the ESD protection structures and the core circuits under protection due to parasitic resistance and capacitance associated with the ESD protection. Such parasitic may be tolerable in IC chips that operate in lower frequency. ESD protection devices for RF applications must hold very low-parasitic capacitance to minimize degradation to RF functionality from poor input/output impedance matching. This requirement poses one of the greatest challenges in RF ESD design, as a low capacitance typically means a small device area and consequently a poor robustness of the ESD-protection device. Other concerns in ESD design for RF ICs include the signal distortion due to the nonlinearity of the parasitic capacitance, the noise coupling as well as noise generation through the ESD protection structures.

In this chapter, various ESD protection options for CMOS RF ICs will be briefly discussed. Then a new SCR-based ESD protection clamp aimed for a very low parasitic capacitance will be presented. This clamp also provides a relatively low-trigger voltage for effectively protecting thin gate oxide as well as a relatively high-holding voltage for latch-up immunity. The capacitance variation across the operation voltage range is small. This linear capacitance is very desirable for RF circuit. The new device can be fabricated in a standard CMOS process without introducing extra processing steps.

Q. Cui et al., *On-Chip Electro-Static Discharge (ESD) Protection For Radio-Frequency Integrated Circuits,* DOI 10.1007/978-3-319-10819-3_2

2.2 ESD Protection Strategy of CMOS RF ICs

In standard CMOS processes, ESD protection circuits can be divided into rail-based clamps and local clamps. The local clamping strategy places clamp structures between I/O pad and supply buses VDD/VSS (Fig. 2.1a) and directly shun negative and positive stress pulses to the grounded power bus (VDD or VSS). The clamps can exist either between the I/O pad and both the upper and lower supply buses (VDD and VSS) or between the I/O pad and one supply bus (VDD or VSS). The rail-based clamping strategy uses two diodes to connect I/O pad and VDD/VSS (Fig. 2.1b). The diodes operate only in forward biased mode and form ESD protection network along with the power clamp.

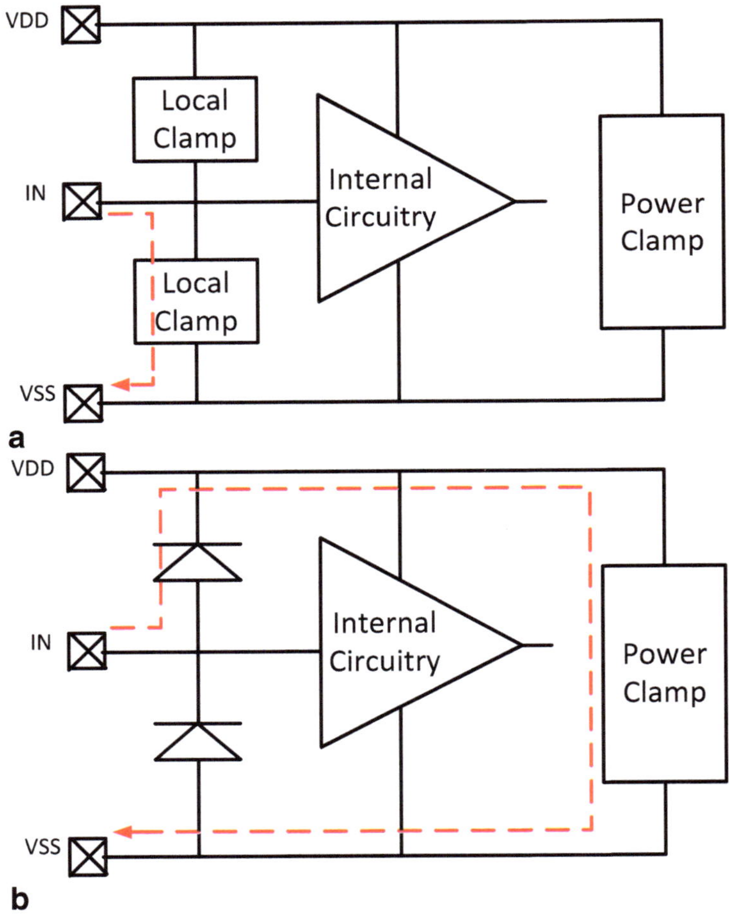

Fig. 2.1 **a** Local clamping ESD protection strategy. **b** Rail-based clamping ESD protection strategy

A simple figure of merit (FOM) may be used to characterize an ESD protection for RF circuits:

$$FOM = Z_{shunt} \cdot V_{HBM} = \frac{V_{HBM}}{\omega C_{ESD}} \tag{2.1}$$

where Z_{shunt} is the impedance of the shunt path due to the parasitic capacitance of the ESD protection, and VHBM is the HBM-ESD pass level. A typical tolerance requirement of return loss is 15 db for transmit/receive circuitry, which is corresponding a shunt impedance of 350 Ω to match a 50 Ω RF source impedance. The maximum loading capacitance for such requirement is about 200 fF, including the parasitic from bond pad itself, for a frequency as low as 2 GHz. Most RF circuits operate at much higher frequency. Therefore, it is a great challenge to minimize the capacitance of protection devices and improve the FOM.

Snapback-type devices, such as ground gate MOS (ggMOS), gate-coupled MOS (gcMOS), silicon-controlled rectifier (SCR), are the commonly used ESD protection devices as local clamps in CMOS technologies. MOS devices are not suitable for RF ESD protection since they are typically large in size and have significant parasitic capacitance. SCR devices can handle much higher current density than MOS structures. Efficient ESD protection has been achieved with various SCR devices in relatively small area, therefore small parasitic capacitance. Two main concerns in using normal SCRs are high-trigger voltage and latch-up issue due to low-holding voltage. A low-voltage-triggered SCR (LVTSCR) structure uses a MOS transistor as a trigger to lower the trigger voltage. Though there is an additional trigger MOS, LVTSCR ESD devices are more compact than the MOS devices with the same protection levels and can have much smaller parasitic capacitance. However, the capacitance of most SCR structures is still too high for many RF applications.

The approach of diodes with power clamp is by far the most widely used ESD protection strategy. While ESD stress is mainly suppressed by the VDD–VSS power clamp, the parasitic capacitance affecting RF performance is mainly from the two ESD diodes which normally can be smaller than the capacitance in local clamps. A large power clamp can be used without significantly impacting the pad capacitance. Snapback-type clamps may be used as the power clamps. Active clamps with large FET can also be used. In the optimization of the ESD diodes, the diode sizes should be large enough to have small on resistance to shun the ESD current but small enough to meet the low capacitance requirement for RF applications. Another factor needs to be considered is that the diodes must have short turn-on time. Generally, poly-bounded diodes turn on faster than regular STI-bounded diodes. Though for the same active dimensions, poly-bounded diodes have higher capacitance than STI-bounded diodes, the latter one have better ESD performance and thus have better FOM values. Stacked diodes in series can be used to replace the single diodes in Fig. 2.1b to obtain lower (total) capacitance. The price is the increased voltage drop across the diodes as well as a large silicon area.

The FOM for ggNMOS, LVTSCR, and diode protections of a 90-nm CMOS process was compared in literature [26]. The ratio of FOMs was 12.5 (diode), 2.7 (ggNMOS), and 5.4 (LVTSCR). It means that the diode solution far better than the ggNMOS and LVTSCR. However, in some RF circuits, the input voltage may swing to higher level than VDD. In such cases, no diode can be placed between the I/O and VDD terminals therefore the rail-based strategy cannot be used. On the other hand, the diode may couple noise from the input to the output through the VDD bus line. As the semiconductor industry moves to more advanced technologies, the bus lines become thinner and longer due to the increased chip size and smaller pitch size. This increases the voltage drop on the bus line and makes the ESD protection less effective, especially in high-current CDM domain. Thus the local clamps can be a more attractive option, even the only option in some cases, for RF ESD protection in CMOS technologies, providing that local clamps with very low capacitance can be developed.

The RF protection design using inductors has been reported in the literature. One configuration uses inductors with power clamp [27]. It is a rail-based clamp in which the diodes are replaced by inductors. It can only have one inductor between the RF I/O pin and VSS and also can have two inductors between the I/O pad and both VDD and VSS. Another configuration combines inductor with capacitor to for a resonator to provide ESD protection [28].

Forward-biased diode chain has traditionally been used as ESD protection structures for RF ICs in CMOS process, as either power clamp or local clamp, because of its relative simple makeup and low-parasitic capacitance [29, 30]. But such a structure can have a large on-state resistance as well as a high leakage current [31]. Significant work has been reported on improving SCRs as ESD clamps for RF applications. A polysilicon SCR was reported in [32] for its low capacitance and good ESD robustness. This ESD clamp is considered a better protection solution than the diode chain, but it requires a modification to the polysilicon deposition process which limits its application in the standard CMOS process. An SCR with a waffle layout and small parasitic capacitance was also developed recently [33]. Nevertheless, this structure exhibited a relatively large turn-on time and large trigger voltage (close to 13 V), rendering it unsuitable for ESD protection of low-voltage CMOS applications. In the next section, the development of a new SCR-based ESD protection clamp aimed for a very low parasitic capacitance will be presented.

2.3 High-Robustness and Low-Capacitance Silicon-Controlled Rectifier for Radio Frequency Protection

2.3.1 Device Structure

Figure 2.2a and b shows the cross-sectional views and equivalent junction capacitances of the conventional low-voltage SCR (Clamp1) and the proposed optimized SCR (Clamp2), respectively. These devices were fabricated in a 0.18-μm CMOS process and have the same layout area of 50×10 μm^2. It is important to note that the

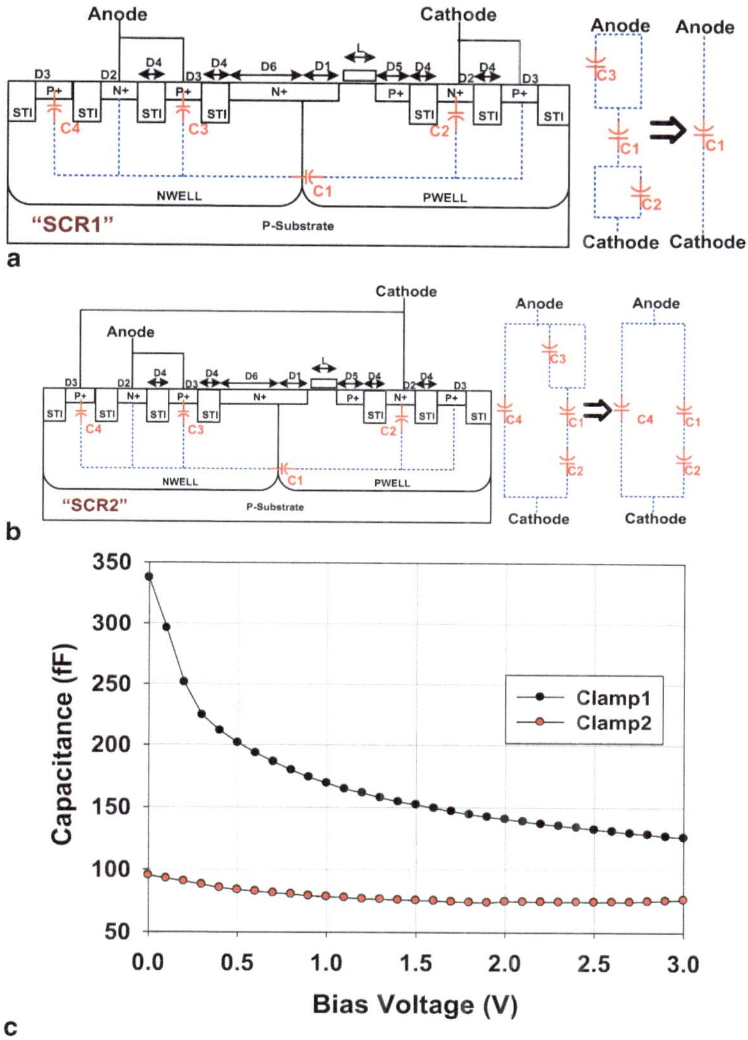

Fig. 2.2 **a** Cross section and capacitance equivalent circuit of conventional-like SCR Clamp1. **b** Cross section and capacitance equivalent circuit of new optimum SCR Clamp2. **c** Measured capacitances versus anode–cathode voltage of Clamp1 and Clamp2

make-up of the two devices is identical, only the cathode terminal connections are different. Instead of connecting the cathode terminal to the P+ region in the P-Well region, as is conventionally done (see Fig. 2.2a), the cathode terminal of the new optimized SCR is connected to the P+ region in the N-Well region (see Fig. 2.2b). Due to this optimum configuration, the total capacitance of Clamp2 is equal to C4 in parallel with C1 in series with C2. On the other hand, the total capacitance in Clamp1 is dominated by C1, since C2 and C3 are shortened.

The aforementioned capacitances are the various junction capacitances imbedded in Clamp1 and Clamp2 shown in Fig. 2.2a and b. Since C1 is the largest among these capacitances due to the large N-Well/P-Well junction area, Clamp2 has a lower capacitance than Clamp1. The capacitances of Clamp1 and Clamp2 at different anode-to-cathode voltages were measured using HP 4284 and are shown in Fig. 2.2c. Notice that the capacitance of Clamp1 is larger than that of Clamp2 by a factor as large as 3.4. The results also show that the capacitance values for both Clamp1 and Clamp2 decrease when the voltage is increased. The trend is consistent with the fact that dominating capacitance C1 associated with the reverse biased N-Well/P-Well junction decreases with the increasing reverse voltage [34]. However, the change of capacitance across the voltage range in Clamp2 is significantly smaller than in Clamp1. This better linearity helps circuit a lot because constant capacitance is much easier to be offset compensated using proper techniques.

2.3.2 Measurements and Discussion

In order to analyze in more detail, the overall parasitic capacitances of Clamp1 and Clamp2, the same SCR-based structure but with isolated anode and cathode terminals was fabricated as shown in Fig. 2.3a. To isolate the various capacitances in the device, several isolated terminals (anode1, anode2, anode3, cathode1, and cathode2) were defined in the various N+ and P+ regions. The different pairings of these terminals allowed capturing the difference in capacitances, as illustrated in the equivalent circuits in Fig. 2.3b. For example, anode2–cathode2 pair yields C1. These different capacitances were measured and plotted in Fig. 2.4.

The results confirm the earlier statement that the main blocking junction capacitance C1 is the largest among all these capacitances. The parasitic capacitance of Clamp1 is about the same as the N-Well/P-Well capacitance C1, which is very large (335 fF at zero bias). In contrast, in the optimized Clamp2, the C1 is in series with C2, thus resulting in a small value of C1,2 (34 fF at zero bias). In addition, the capacitance of P+/N-Well junction C4 is also small (60 fF at zero bias) because of the small junction area. Hence the parasitic capacitance of Clamp2 is simply the

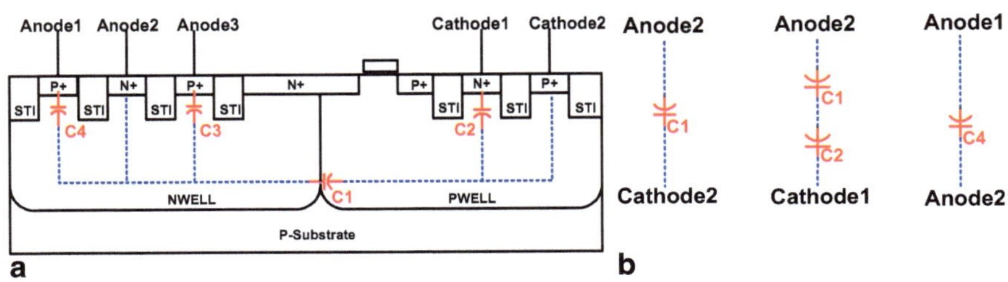

Fig. 2.3 **a** SCR with isolated anode and cathode terminals. **b** Equivalent capacitances for the pairings of anode2–cathode2, anode2–cathode1, and anode1–anode2

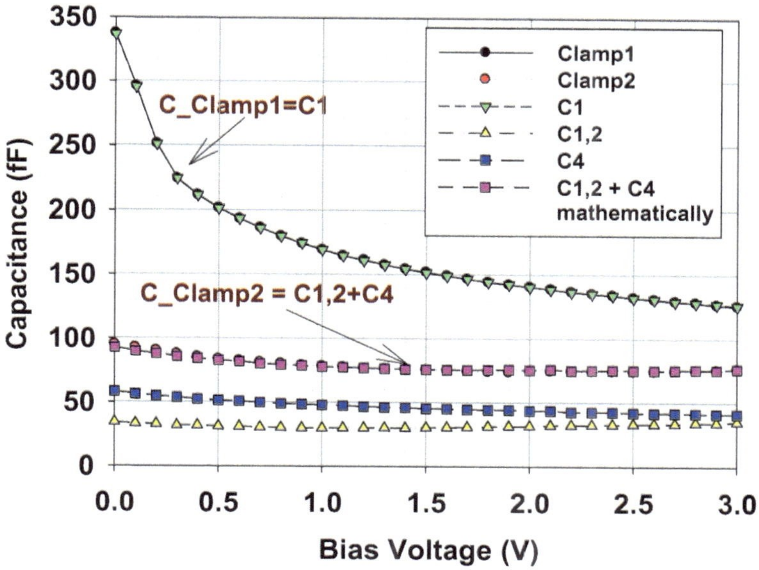

Fig. 2.4 Comparison of capacitances obtained from different anode–cathode connections shown in Fig. 2.3

addition of C4 and C1, 2 (see Fig. 2.4), and Clamp2 exhibits a much lower capacitance than Clamp1 (94 fF versus 335 fF).

To investigate the ESD robustness of Clamp1 and Clamp2 under the human body model (HBM) stress, transmission line-pulsing (TLP) measurements were conducted. Poststress leakage currents were measured at 3.6 V, which is the maximum operating voltage of the technology under consideration. Figure 2.5 shows the measured TLP current (y-axis) versus voltage (bottom x-axis) and leakage current (top x-axis) curves of Clamp1 and Clamp2. In the forward direction, Clamp2 has a smaller trigger voltage and higher failure current than Clamp1 (i.e., 8.1 V in Clamp2 versus 10.5 V in Clamp1 and 1.55 A in Clamp2 versus 1.2 A in Clamp1). These improved features are stemmed from the presence of the floating base of the N-Well/P-Well/N+ transistor in Clamp2. In the reverse direction, Clamp2 also possesses a smaller on-state resistance and higher failure current than Clamp1. This is owing to the different current conducting capabilities of the P-Well/N-Well diode in Clamp1 and the P+/N-Well diode in Clamp2.

Table 2.1 summarizes the characteristics of the two above-mentioned ESD devices (Clamp1 and Clamp2) as well as another low-capacitance clamp (Clamp3) reported previously [35]. Among the three devices, Clamp2 is the best candidate for RF ESD protection because of its all-around excellence in parasitic capacitance, current handling capability, trigger voltage, turn-on time, and voltage overshoot. It should be pointed out that the turn-on time and voltage overshoot were measured using the very-fast TLP and are important parameters for a fast ESD event like the charge device model (CDM). The voltage overshoot is the peak voltage of ESD

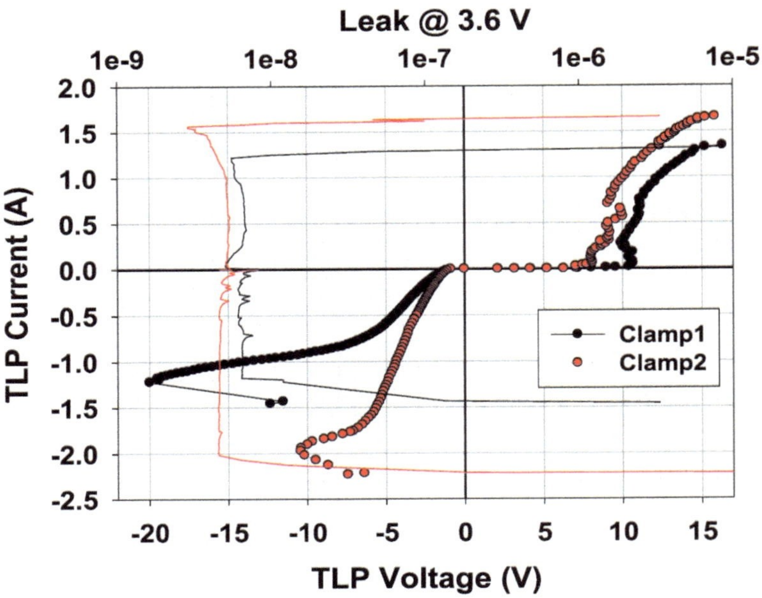

Fig. 2.5 Bidirectional TLP current–voltage characteristics (*red and black closed circles*) and leakage currents (*red and black solid lines*) of Clamp1 and Clamp2

Table 2.1 Comparison of results of Clamp1, Clamp2, and Clamp3

Parameter	Clamp1	Clamp2	Clamp3 [35]
Device Area (μm^2)	50 × 10	50 × 10	45 × 20
Failure current-forward (A)	1.2	1.55	1.0
On-state resistance-forward (Ω)	5.0	4.5	4.8
Trigger voltage-forward (V)	10.5	8.1	8.5
Failure current-reverse (A)	1.2	2.0	1.5
On-state resistance-reverse (Ω)	13.0	2.0	2.5
Turn-on time (ns) @ 1 A	3.2	0.9	2.8
Voltage overshoot (V) @ 1 A	25	12	22
Capacitance @ 0 V bias (fF)	335	94	95

device in its turn-on transient, and the turn-on time is defined as the time from voltage overshoot point to 110 % holding voltage point.

The typical capacitance of a ggNMOS for 2 kV HBM protection is about ~ 500–700 fF. The diode network in a rail-based clamp with the same ESD protection level has a typical capacitance around 100 fF. The failure current of the new clamp is 1.55 A which is equivalent to 2.3 kV HBM failure level. Thus the new clamp provides a desirable local clamp that has a better FOM of RF ESD protection than a typical rail-based diode network clamp.

2.4 Summary

In this chapter, various ESD protection options for CMOS RF ICs have been discussed. The local clamping strategy with low-capacitance clamps is a preferred RF ESD protection scheme. A new SCR-based ESD protection clamp device with a very low parasitic capacitance has been proposed and fabricated in a standard CMOS process without introducing extra processing steps. The new ESD structure has been characterized experimentally for the design of effective ESD protection solutions in high-frequency CMOS-based ICs. Such a new device can offer a relatively low-trigger voltage for effectively protecting thin gate oxide and a high-holding voltage for latch-up immunity. It also demonstrated the smallest peak voltage and fastest response time. The device showed a much lower parasitic capacitance, a better capacitance linear, and better ESD performance than the conventional SCR and a low-capacitance SCR reported in the literature. A theoretical analysis was also given to provide the physics underlying the low capacitance observed in the new ESD protection device.

Chapter 3
Design of SiGe SCR Devices for Radio Frequency Integrated Circuits in SiGe BiCMOS Process

3.1 Introduction of SCR Device Technology

Silicon controlled rectifiers (SCRs) are increasingly used in the industry for electrostatic discharge (ESD) protection applications because of their high current handling capability [36] and low parasitic capacitance [37, 38]. In particular, these devices continue gaining importance for constructing ESD protection solutions for modern high-frequency integrated circuits such as those fabricated in a SiGe BiCMOS process [39].

ESD protection solutions for SiGe-based integrated circuits have been traditionally built primarily using low capacitance diodes. For applications in which the operating conditions do not allow for the use of up/down diodes due to voltage swing or noise considerations, Si-based low-voltage trigger SCR (LVTSCR) has also been considered [29, 32]. While the LVTSCR is generally robust, its response time is relatively slow. The response time for ESD device is defined as the time it takes the device to sink the intended current forced through it. LVTSCR's "slow" response performance leads to a considerable transient voltage overshoot that can severely limit the capability of this device to protect the MOS gates connected to the I/O pins during fast transient events, such as the charged device model (CDM) event. In light of this, an improved SCR with a dummy gate structure has been also proposed to improve the response time and reduce parasitic capacitance [40].

In an effort to achieve a more robust protection clamp, SiGe SCRs characterized via transmission line pulsing (TLP) testing have been reported in the literature [2, 39, 41, 42]. These studies provided SiGe SCR's quasi-static I–V behaviors, which are applicable for the human body model (HBM) ESD event. However, information pertinent to SiGe SCR's transient behaviors, such as the peak voltage and response time, has not been addressed. The transient behaviors are less critical for optimizing the protection device for HBM robustness, but gain importance for achieving CDM protection robustness. CDM is a fast ESD event having pulses with a 100–200 ps rise time and 1–2 ns width. More recently, the SiGe SCR's transient performance under the CDM stress was reported in [43], which offered initial guidelines for optimizing SiGe SCRs.

Q. Cui et al., *On-Chip Electro-Static Discharge (ESD) Protection For Radio-Frequency Integrated Circuits,* DOI 10.1007/978-3-319-10819-3_3

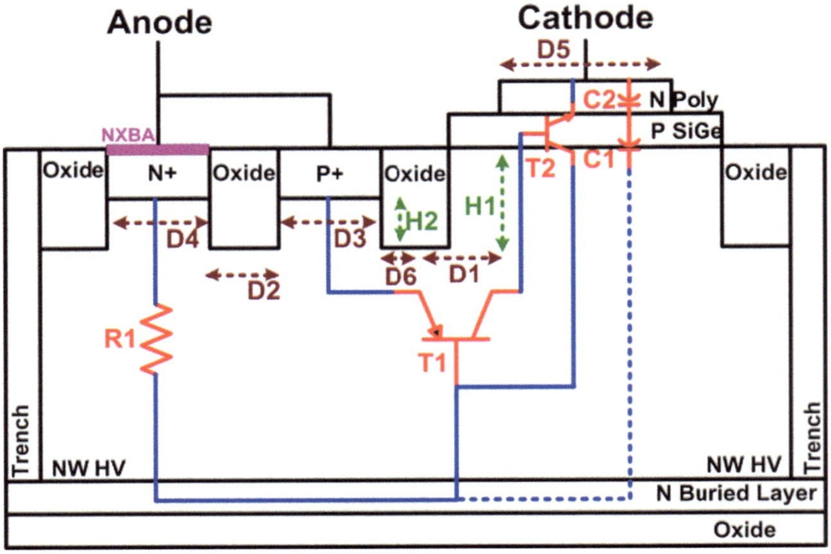

Fig. 3.1 Cross-section views and equivalent circuits of conventional SiGe SCR (SCR1)

This chapter applies a systematic design methodology for SiGe BiCMOS SCR clamps based on ESD technology computer aided (TCAD) analysis [44–50]. Very fast transmission line pulsing (vfTLP) measurements and TCAD simulations are conducted to obtain the underlying understanding on the optimization of several SiGe SCRs subject to the CDM ESD event and optimization results are demonstrated via measurements and application examples.

A conventional SiGe silicon controlled rectifier (SCR1) [2] built in a 0.18-μm SiGe BiCMOS process is shown in Fig. 3.1, including its cross-section view and equivalent circuit. SCR1 consists of a P+/NW HV/P bipolar transistor T1, NW HV/P SiGe/N Poly bipolar transistor T2, NW HV resistor R1, and main capacitances C1 and C2. Note that there is an underneath buried oxide layer in this process and a deep trench surrounding this device. And also there is an additional highly doped NXBA layer between N+ and anode metal that helps a lot to reduce the contact resistance. On the contrary, no such highly doped layer is available to P+ contact for SCR1 structure due to process limitation. Therefore some contact resistance effect might be expected here. Unlike T1, which is a Si homojunction bipolar transistor, T2 is a heterojunction bipolar transistor that turns on much faster than T1. In such a device, T1's base width D1 (i.e., 0.7 μm) is designed to be the minimum value allowed by the design rule of this process, the layout area is 208 μm^2 and the device width is 50 μm.

ESD measurements were carried out using the Barth 4012 vfTLP tester. It generates CDM-like pulses with a 0.1 ns rise time and 5 ns width [46, 47, 51]. Post-stress leakage currents were measured at a voltage of 3.6 V.

Figure 3.2 shows the transient voltage and current waveforms, respectively, under a vfTLP pulse which generates 570 mA effective DUT (device under test)

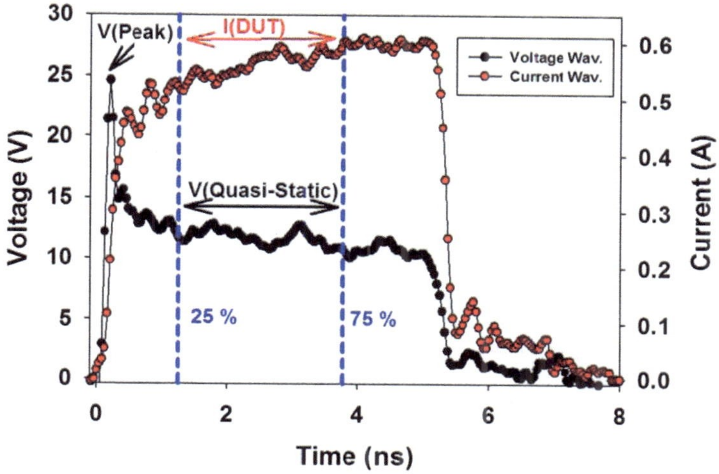

Fig. 3.2 Measured transient waveforms for "SCR1" under Barth 4012 vfTLP with 5 ns pulse width, 0.1 ns rise time at 570 mA DUT current level stress: Voltage waveform (*looking to left*) and current waveform (*looking to right*)

current level. Note that the 570 mA effective vfTLP current level was used as a benchmark and corresponds to an arbitrary point in the snapback region of operation of the SCRs (see Fig. 3.3). It can be seen that the transient voltage waveform first rises to its overshoot peak value (~24.61 V at around 0.1 ns) and then falls to

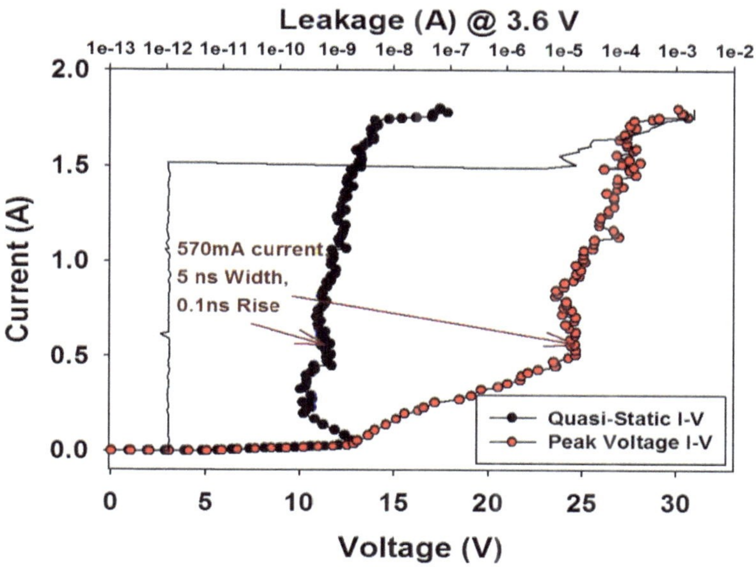

Fig. 3.3 Measured I–V curve in Barth 4012 vfTLP for SiGe SCR1 using the quasi-static and peak voltage methods

a relatively "flat" region (~ 1 ns). On the other hand, the current exhibits a plateau-like waveform between the turn-on and turn-off window. Based on these results, both the quasi-static method V(Quasi-Static) and peak voltage method V(Peak) can be used to extract the voltage data. For the quasi-static method, the data are extracted by averaging the voltage waveform in the 25–75 % window, whereas the peak voltage method simply takes the peak transient voltage. On the other hand, only the quasi-static method is used in extracting the current data from the waveform since the overshoot is absent, and this extracted current data will be plotted against extracted quasi-static voltage and peak voltage respectively in the followings.

Figure 3.3 compares SCR1's vfTLP current (y-axis) vs. voltage (bottom x-axis) and leakage current (top x-axis) curves using the quasi-static voltage and peak voltage methods. Clearly, the I–V curves of the two methods diverge soon after the device turns on at around 12.8 V. For example, at a current level of 570 mA, the quasi-static voltage is 11.62 V but a much higher peak voltage of 24.61 V (see Fig. 3.3). This phenomenon indicates that the commonly used quasi-static I–V method can overestimate ESD device's protection ability under fast ESD events such as the CDM. From the viewpoint of the quasi-static method, SCR1 looks good enough since the clamping voltage is below 13 V until it fails at 1.5 A. Based on the peak voltage method, this device in fact is poor because its peak voltage keeps on increasing to more than 25 V, a sufficiently large stress to damage the internal circuits intended to be protected (e.g., thin gate oxide).

In the following, we will study SiGe SCRs based on vfTLP measurements and TCAD simulations to gain the physical insights on the transient behaviors (peak voltage and response time) under the CDM event.

3.2 An Improved Vertical SCR-Based ESD Protection Design

3.2.1 Device Structure

Aiming at reducing the peak voltage, a new SiGe SCR with a P PLUG layer added was proposed and demonstrated recently [43]. For this specific process, a highly-doped PXBA layer was an associated layer of PPLUG and being automatically added between anode metal and N+ layer, which further reduces the contact resistance. The cross-section view and equivalent circuit of such a device is shown in Fig. 3.4. Here we consider two different versions, called SCR2 and SCR3. They consist of a P PLUG/NW HV/P bipolar transistor T3, NW-HV/P SiGe/N Poly bipolar transistor T4, NW-HV resistor R3, and main capacitances C3 and C4. For SCR2, the dimension D1 is set to the minimum value of 0.7 μm, and D1 is enlarged to 1.7 μm in SCR3. The areas are 208 and 258 μm^2 for SCR2 and SCR3, respectively. Both devices have the same width of 50 μm.

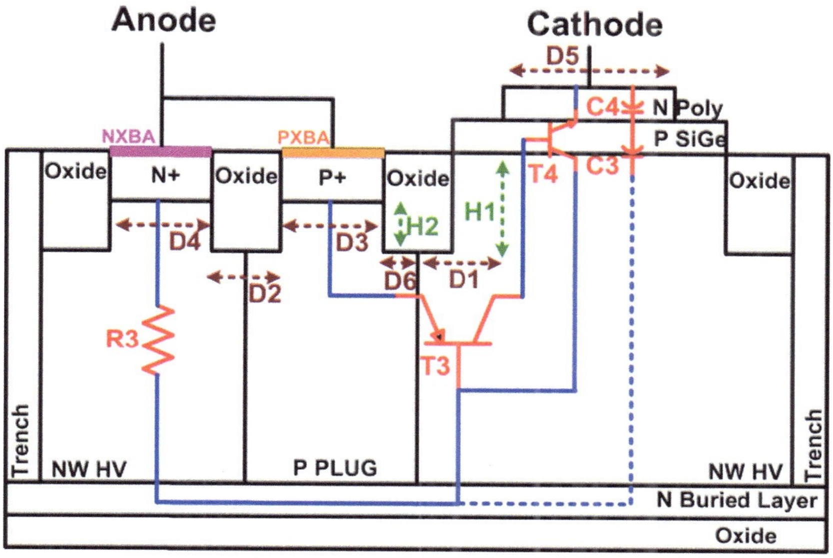

Fig. 3.4 Cross-section views and equivalent circuits of proposed SiGe SCRs with P PLUG (SCR2 and SCR3)

3.2.2 Measurements and Discussion

Figure 3.5a compares the transient voltage waveforms of SCR1, SCR2, and SCR3 under a 570 mA effective current level vfTLP stress. The results demonstrate that the presence of P PLUG can reduce the peak voltage, as SCR1 has the largest peak voltage of 24.61 V and SCR2 and SCR3 have smaller peak voltages of 13.91 V and 15.65 V. PPLUG can also reduce the overshoot magnitude defined by difference of the peak voltage "V(Peak)" and averaged holding voltage "Vave" as SCR1 has 13.1 V overshoot magnitude and SCR2 and SCR3 only have 11.0 V, 12.7 V respectively. It is noticed that SCR1 has the largest holding voltage "Vave" (averaged value between 25 and 75 % time window), which mainly come from contact resistance of anode metal and P+ in Fig. 3.1. The metal/P+ contact resistance effect should not hurt our analysis of SiGe SCR under CDM condition since both measurement and simulation results show that turn-on speed indicators (peak voltage and overshoot magnitude) are not affected by contact resistance. The current waveforms of the three devices are given in Fig. 3.5b.

Figure 3.6 shows the I–V curves of SCR1, SCR2, and SCR3 extracted using the peak voltage method. Consistent with the results in Fig. 3.5a, SCR2 is clearly the most optimized ESD device in terms of the maximum voltage across it during vfTLP stress. It is interesting that from Fig. 3.6, after Vt1 points, changes of I–V slopes around 0.5 A can be observed on all three SCRs (around 15 V for SCR2

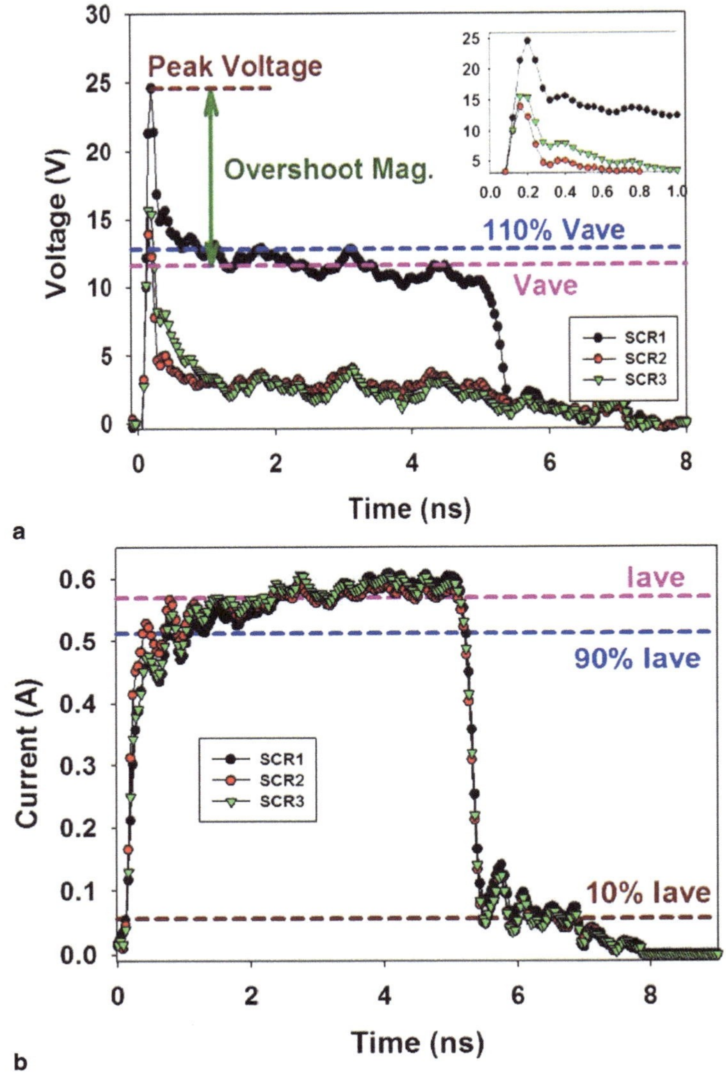

Fig. 3.5 Comparison of measured transient **a** voltage waveforms and **b** current waveforms for SiGe SCRs (SCR1, SCR2, and SCR3) subject to 570 mA effective DUT current level with 5 ns pulse width, 0.1 ns rise time by Barth 4012 vfTLP stress

and SCR3, and 25 V for SCR1) under peak voltage extraction method. This slope-changing phenomenon confirms with the statement raised in [45, 52, 53] that turn-on time decrease with increased vfTLP current level. The explanation is as follows: when vfTLP current level is lower than certain value (e.g., <0.5 A, region "I"), the displacement current (dE/dt) is unable to sufficiently trigger the turn-on process at ~0.1 ns time point, therefore the peak voltage continues to increase (region "I" in

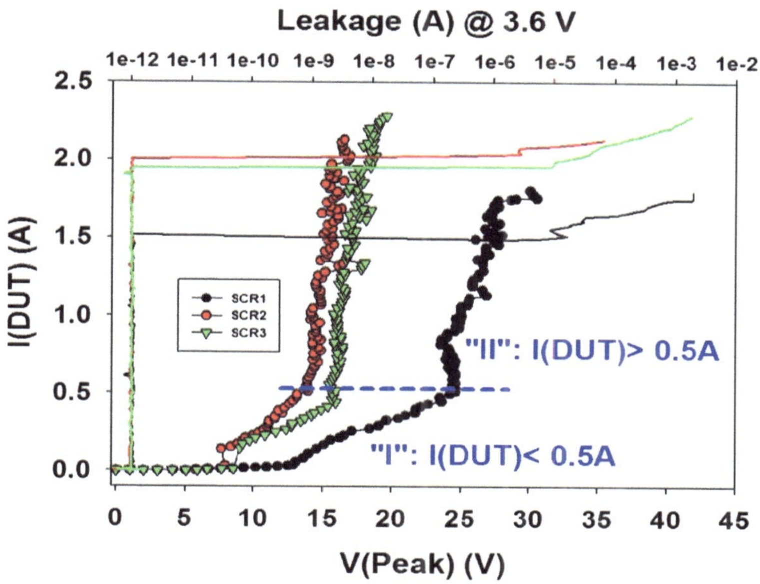

Fig. 3.6 Comparison of I–V curves extracted using the peak voltage method for SCR1, SCR2, and SCR3

Fig. 3.6). While as the vfTLP current is greater (e.g., >0.5 A, region "II"), the turn on process was triggered at 0.1 ns time period, which limits the peak voltage and exhibits a steeper I–V slope. This slope-changing phenomenon cannot be observed by traditional quasi-static voltage extraction method because the traditional quasi-statci method only takes data from 25 to 75 % time window when the device has already been turned on. The 570 mA vfTLP current level mentioned above was intentionally selected from region II to better compare the turn-on performance of different versions of SiGe SCRs.

Another important figure of merit for the CDM event is the SCR's turn-on response time, which can be extracted using both voltage vs. time waveforms and the current vs. time waveforms given in Fig. 3.5a and b. As for voltage turn-on time method, response time is defined by the time from peak voltage to 110 % averaged holding voltage "Vave." For all the three SiGe SCRs considered, the currents increase rapidly before 0.2 ns and then reach a fairly constant level. The time for ESD device to reach its steady-state current is called the current turn-on response time and is defined as the time it takes from 10 to 90 % Iave, where Iave is the constant current calculated by averaging the currents between 25 and 75 % time periods. The values of Iave, 10 % Iave, and 90 % Iave for SCR1 are labeled in Fig. 3.5b. For both voltage turn-on method and current turn-on method, SCR2 has smallest turn-on response time, SCR3 has the largest response time, and SCR1 has a value in between. For constructing an effective CDM ESD protection solution, a small peak voltage and small response time are highly desirable. Thus, SCR2 with the P PLUG layer and minimum D1 is the most optimized device for such an application.

Table 3.1 Summary of dimensions and measurement ESD parameters

Device	SCR1	SCR2	SCR3
Figures	Fig. 3.1	Fig. 3.4	Fig. 3.4
D1 (μm)	0.7	0.7	1.7
D2 (μm)	1.0	1.0	1.0
D3 (μm)	0.7	0.7	0.7
D4 (μm)	0.7	0.7	0.7
D5 (μm)	0.35	0.35	0.35
D6 (μm)	0.5	0.5	0.5
H1 (μm)	0.5	0.5	0.5
H2 (μm)	0.2	0.2	0.2
P PLUG existence	N	Y	Y
Effective PNP base width (μm)	1.9	1.2	2.2
Device Width (μm)	50	50	50
Trigger (V)-vfTLP	12.77	6.33	8.16
Trigger (V)- TLP	12.33	7.1	8.03
Failure current "I_{t2}" (A)-vfTLP	1.5	2.0	1.94
Failure current "I_{t2}" (A)-TLP	0.7	1.5	1.4
On-state resistance (Ohm)-vfTLP	18.49	8.44	9.01
On-state resistance (Ohm)-TLP	21.31	3.25	3.61
Holding voltage (V)-vfTLP	11.62	3.0	2.5
Holding voltage (V)-TLP	8.0	3.0	3.0
Peak voltage (V) @ 570 mA	24.61	13.91	15.65
Overshoot mag. (V) @ 570 mA	13.1	11.0	12.7
Voltage turn-on time (ns)@570 mA	0.96	0.5	1.02
Current turn-on time (ns)@570 mA	0.63	0.27	0.64

Table 3.1 summarizes the dimensions and measured key parameters of the three SCR's under study using both vfTLP measurement and traditional TLP measurement. It is worthy to mention that both vfTLP/TLP measurements show exactly the same tendency for parameters such as trigger voltage, failure current, on-state resistance, and holding voltage. For this specific process, the trigger voltage, peak voltage of ESD protection device under CDM ESD event is targeted to be lower than 14 V, which is the gate oxide breakdown voltage in this process. The SiGe SCRs mentioned in this chapter were fabricated in a dielectrically isolated SOI process, so there is inherent latch up immunity. And the quasi-static holding voltage is targeted to be larger than operation voltage "1.8 V-Vdd" to further avoid latch up risk. At the same time, high failure current is favorable. Clearly, the SCR2 is the most optimized ESD device because of its all-around excellence in the parameters mentioned before. And as the fabrication technology further scales down (e.g., 90 nm, 65 nm), the proposed SCR2 will turn out to be more advantageous for its smaller overshoot/faster turn-on speed because in such advanced process, gate oxide is very thin and therefore more and more vulnerable under CDM stress. In the next section, we will

Table 3.2 Doping profile in TCAD simulation

Layer	Doping	Concentration (cm^{-3})
P SiGe	Boron	1.2E17
N Poly	Arsenic	5E20
NW HV	Phosphorus	1E16
P PLUG	Boron	1E17
N Buried	Arsenic	1E19
N+	Arsenic	1E19
P+	Boron	1E19

conduct TCAD simulations to analyze and gain physical insights on the transient behaviors (peak voltage and response time) of the SCRs under the CDM event.

The SiGe SCRs were simulated in Sentaurus TCAD software tool using the same doping densities listed in Table 3.2. The TCAD simulation setup was calibrated by comparing DC-like I–V curves between simulation and measurements. We performed vfTLP-like simulations in which voltage pulses with a 0.1 ns rise time, 5 ns width and incremental amplitude were used. An effective resistor was also attached to the voltage pulse to mimic 50 Ω transmission impedance of vfTLP test system. In this TCAD simulation, important mechanisms including the Fermi statistics, Shockley-Read-Hall carrier recombination, Auger recombination, high-field saturation, lattice temperature, and avalanche model are accounted for.

The simulated voltage vs. time waveforms for SCR1, SCR2, and SCR3 are shown in Fig. 3.7. All devices exhibit a peak voltage at ~0.1 ns before they settle

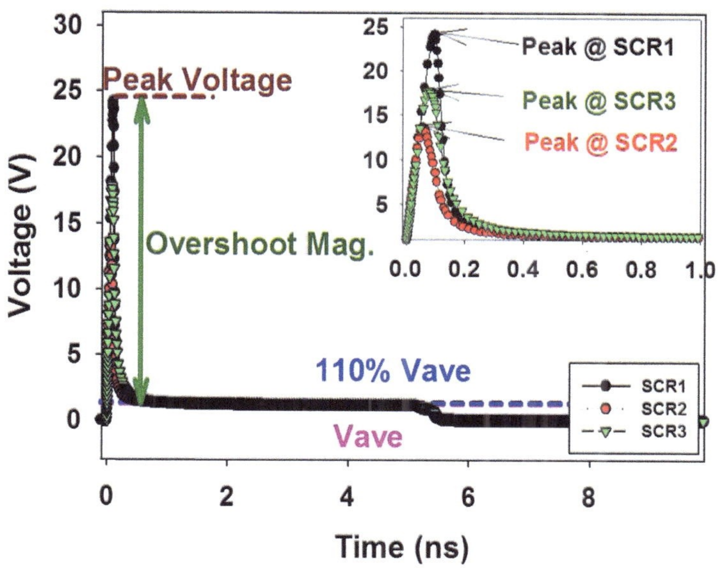

Fig. 3.7 Comparison of simulated transient voltage waveforms for SiGe SCRs under 30 V amplitude, 5 ns pulse width, 0.1 ns rise pulse

down to a relatively constant value. Further, the results suggest that SCR1 has the largest peak voltage and SCR2 the smallest peak voltage. These TCAD simulation data are in excellent agreement with the measured data presented in the previous section except for the holding voltage, because the anode metal/P+ was set to be the ideal contact in simulation hence SCR1 shows no higher holding voltage as seen in Fig. 3.5a. This discrepancy between simulation and measurements should not hurt the physical understanding since peak voltage and overshoot magnitude turn out to be exactly the same tendency as measurements.

In Fig. 3.8, we compare the impact ionization rates of the three SiGe SCRs at their respective peak voltage points (see "Peak @ SCR1," "Peak @ SCR2," and "Peak @ SCR3" in Fig. 3.7). Based on this, we can suggest the following sequential events in the SCRs at the time of voltage peaking [45, 48]:

i. As voltage between the anode and cathode increases, a large number of hole-electron pairs are generated in the high impact ionization region (impact region) between the P+ and P SiGe regions;
ii. Holes are injected toward the cathode to turn on the vertical NPN BJT (see T2 in Fig. 3.1 and T4 in Fig. 3.4) and electrons are injected toward the anode to turn on the PNP BJT (see T1 in Fig. 3.1and T3 in Fig. 3.4);
iii. When both the PNP and NPN BJT are turned on, a large current flows through the impact region; and
iv. Voltage between the anode and cathode decreases.

In other words, the peak voltage is the voltage needed to generate the required impact ionization rate in the impact region to trigger the SCR. Figure 3.8a–c clearly shows that the sizes of the impact ionization region of the SCRs are different, the SCR1 being the largest and SCR2 the smallest. Therefore, the reason SCR1 has the largest peak voltage is because the absence of the P PLUG layer induces a relatively large area of impact ionization region, and the voltage drop roughly equals the integration of electric field along with ionization region. Therefore, larger-area impact ionization region means larger peak voltage needed to trigger the SCR device. On the other hand, for the SiGe SCRs with the P PLUG layer (SCR2 and SCR3) the impact ionization region is much smaller, which in turn reduces the voltage required to generate the required impact ionization. Another factor that influences the area size of the impact ionization is the emitter injection efficiency of the PNP (T1). For SCR2 and SCR3 the presence of the heavily doped PPLUG region will result in a very high emitter injection efficiency factor. The reason is the reduced base width of PNP transistor that improves bipolar efficiency of PNP and eventually improved SCR's turn-on behavior. Whereas for SCR1 this factor is much lower due to the absence of the PPLUG.

Figure 3.9 compares the simulated current vs. time waveforms for SCR1, SCR2, and SCR3 subjected to the same CDM stress as used in Fig. 3.5. A similar trend is found between the measured and simulated data; SCR2 has the smallest response

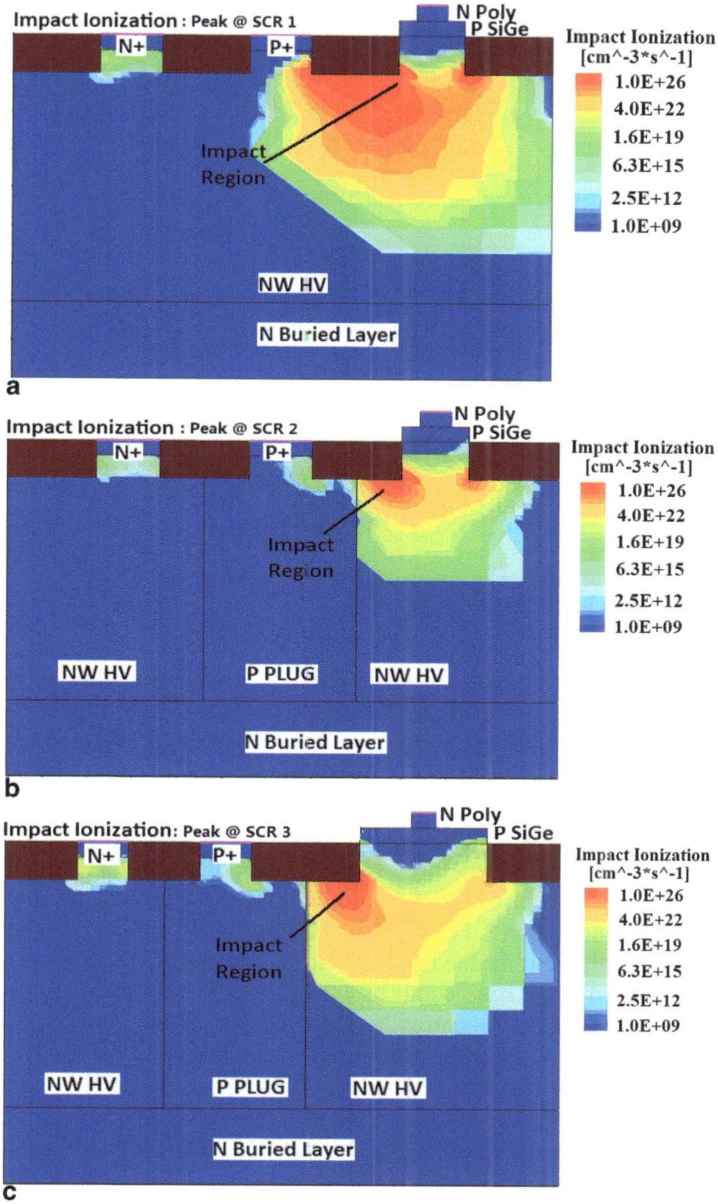

Fig. 3.8 Simulated electron impact ionization contour at peak voltage points of **a** SCR1, **b** SCR2, and **c** SCR3

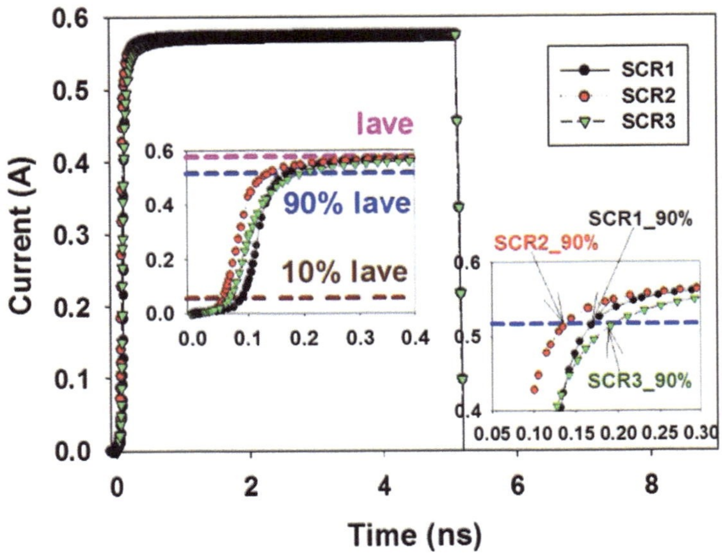

Fig. 3.9 Comparison of simulated transient current waveforms for SiGe SCRs under 30 V amplitude, 5 ns pulse width, 0.1 ns rise pulse

time and SCR3 has the largest response time. To understand this, we simulated the hole current densities of the three SiGe SCRs at their respective turn-on points (see "SCR1_90 %," "SCR2_90 %" and "SCR3_90 %" in Fig. 3.9), and the simulation results are given in Fig. 3.10a–c. The response time of SiGe SCR is mainly determined by the base transit time of the parasitic PNP homojunction transistor (T1 in Fig. 3.1 and T3 in Fig. 3.4) since the vertical NPN transistor (T2 and T4) are heterojunction transistors which turn on much faster than the homojunction counterpart [54]. Thus, a longer PNP base width will increase the base transit time and lead to a larger response time. The effective base widths in which the majority hole currents flow are different for the three SCRs. As indicated in Figs. 3.10a–c, SCR2 has the smallest effective base width of 1.2 μm and SCR3 has the largest effective base width of 2.2 μm. This provides a good explanation to the response times extracted from the measured and simulated results. Comparing the results of SCR1 and SCR2, which have the same D1 value, it can be concluded that the presence of the P PLUG layer reduces the effective PNP base width. To a smaller extent, the smaller D1 in SCR2, in comparison with that in SCR3, also plays a role in reducing the response time of the SiGe SCR.

Table 3.3 summarizes SiGe SCRs' transient parameters obtained from the measurements and TCAD simulations. The preceding analysis has suggested that a SiGe SCR having a combination of the P PLUG layer and minimal PNP base width D1 can offer an optimal protection capability against the CDM ESD event.

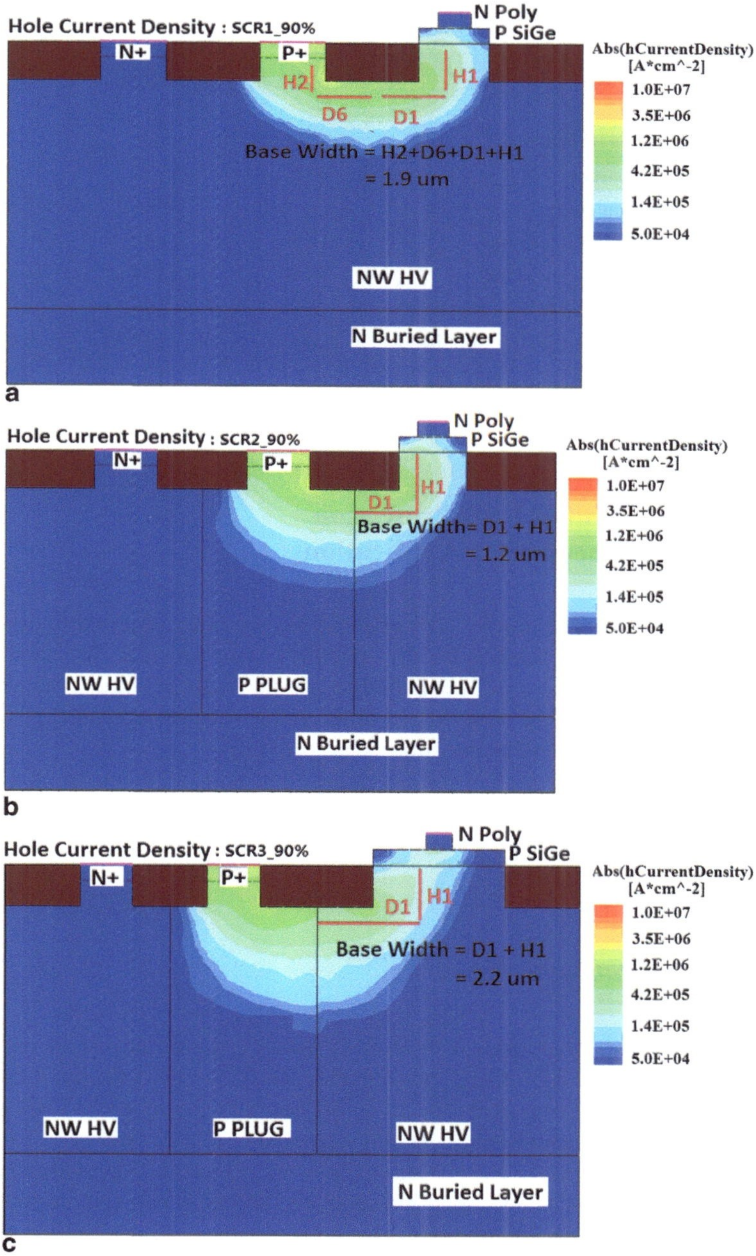

Fig. 3.10 Simulated hole current density contours of **a** SCR1, **b** SCR2, and **c** SCR3

Table 3.3 Summary of transient parameters obtained from measurement and simulation under vfTLP stress

Device	SCR1	SCR2	SCR3
P PLUG Existence	N	Y	Y
Effective PNP base width (μm)	1.9	1.2	2.2
Peak voltage (V): measured	24.61	13.91	15.65
Peak voltage (V): simulated	24.16	13.64	17.82
Overshoot magnitude (V): measured	13.1	11.0	12.7
Overshoot magnitude (V): simulated	22.88	12.36	16.49
Voltage turn-on response time (ns): measured	0.96	0.5	1.02
Voltage turn-on response time (ns): simulated	0.63	0.59	0.92
Current turn-on responstime (ns): measured	0.63	0.27	0.64
Current turn-on response time (ns): simulated	0.09	0.06	0.13

3.3 Optimization of ESD Protection for High Speed SiGe Applications

High frequency integrated circuit (IC) applications are often required to operate at variety of conditions, depending on the specific circuit function. In fundamental terms, the circuit interface can be required to handle small or large signal, narrow band or wide band. In line with the specific circuit requirements, the protection architecture selection would be constrained accordingly.

Achieving satisfactory high speed operating conditions while ensuring a robust level of ESD immunity in circuit applications requires a set of steps in the design process, extending from device-level characterization to systematic codesign for on-chip protection integration.

3.3.1 Capacitance Analysis

Loading capacitance associated with an ESD protection structure is an important consideration in the design of high performance RF IC's. The definition of a FOM (figure of merit) helps in making the trade-off between the required ESD protection level and circuit performance. Loading capacitance at zero-bias per 1 kV HBM classification test level is the FOM used in this work. Figure 3.11 provides the C–V characteristics of the SCR described in this work. The SCR can sink current up to 2.1 A when tested using a 100 ns pulse. This corresponds to a 3.0 kV HBM classification test level. The SCR therefore has a FOM@0 V = 15 fF/1 kV HBM.

In order to analyze the overall capacitance in greater detail it is important to understand the capacitance contributions of the various junctions. Figure 3.12 (redrawn from Fig. 3.4) shows a simplified schematic depicting the various junction capacitances. From the cross section shown in Fig. 3.4 it is observed that the N-EPI

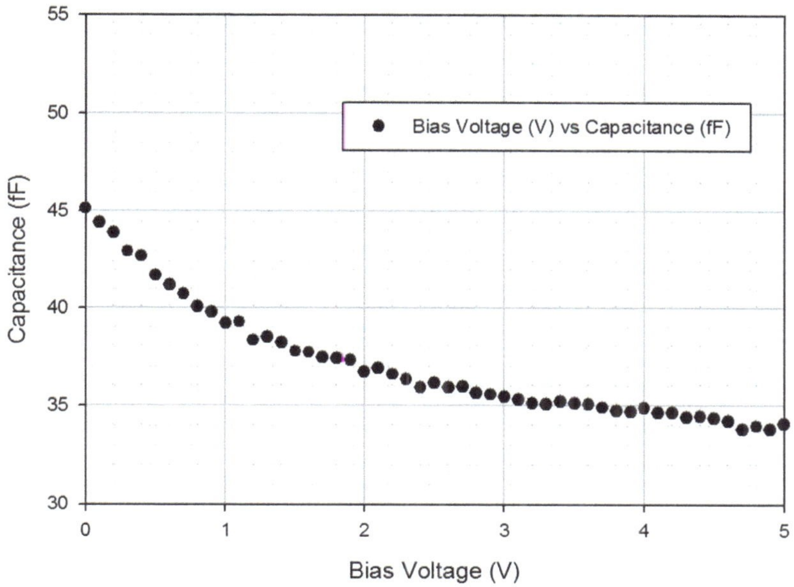

Fig. 3.11 Measured C–V characteristics

Fig. 3.12 Vertical SiGe SCR schematic

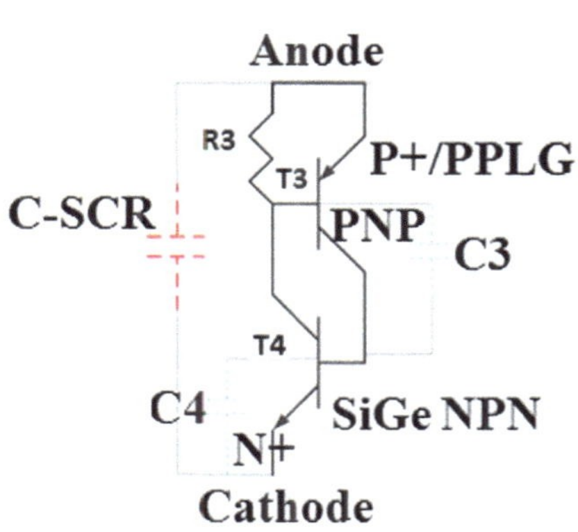

Well/P-SiGe junction (C3) occupies the largest area. Apart from this, we found another capacitance C4 that is associated with the N+/P-SiGe junction of the NPN. In the SCR architecture being proposed, the base of the SiGe NPN is floating. Thus the overall capacitance of the SCR (C-SCR) is the series combination of C3 and C4. Since C4 is a much smaller junction than C3, it dominates and helps reduce the overall capacitance of the SCR. Thus the floating NPN not only helps to reduce the trigger voltage but also results in reducing the overall capacitance associated with the SCR [9]. Metallization layout is also a major contributor of ESD device's capacitance. It changes with process technology and recommended guidelines should be practiced to minimize the extra parasitic capacitance impact.

3.3.2 ESD Devices for On-chip RF Integration

For high speed circuit application that have large signal swing, the ESD protection can be referenced to ground to mitigate risk of noise coupling from and to the power supply domains, which negatively impacts the signal integrity at the I/Os. In these cases, combination of the SCR, discussed earlier in this chapter, and diodes can be considered for the flexible blocking voltage protection implementation. Figure 3.13a, b shows the cross section of complementary SiGe SCR with (a) series p-diode (b) and series n-diode, respectively. In this case, the device formation in Fig. 3.13a provides the SCR blocking voltage plus two p-diodes blocking voltage for RF I/O to VSS protection, while the structure in Fig. 3.13b is used in parallel to provide symmetry below VSS blocking voltage for protection for negative stress at the RF I/O vs. VSS. In this particular example, the DT (Deep Trench) isolation also aids to improve the RF/Microwave performance of the ESD cell, by reducing the lateral parasitic capacitance to substrate.

3.3.3 ESD Protection in RF PADS Configuration: Schematic Applications

The example ESD protection configured in standard GSG for microwave I/Os applications is shown in Fig. 3.14. The package to bond pad transition typically changes for every product application. As part of the design and product development, S-Parameter models would need to be generated accordingly to account for the corresponding pads effect in an accurate model of the impedance transition to the core circuit.

Another typical protection strategy used for narrow-band and wide-band applications also takes advantage of the properties of passive components over frequency [55–58]. As a simple example to illustrate this concept, an ESD protection strategy for microwave I/O's is shown in Fig. 3.15 for narrow-band applications. An inductor is utilized as a shunting element for the ESD current. It becomes a part of the matching network in conjunction with the 50-Ω termination resistance [59]. This protection solution has a much lower impact on the S11/S21 performance of the circuit. This configuration has been demonstrated in LNA designs operating in frequency bands between 6 and 26 GHz.

3.3.4 ESD Protection in Reference RF Circuit Schemes Using SiGe SCR

Figure 3.16 shows an alternative ESD protection scheme for an I/O where a protection to VDD is avoided. Notice the SCR here used similar concept discussed in Fig. 3.12 to lower the parasitic capacitance. This type of protection is often considered for high frequency clock IOs, reducing the effect of noise coupling from the VDD power.

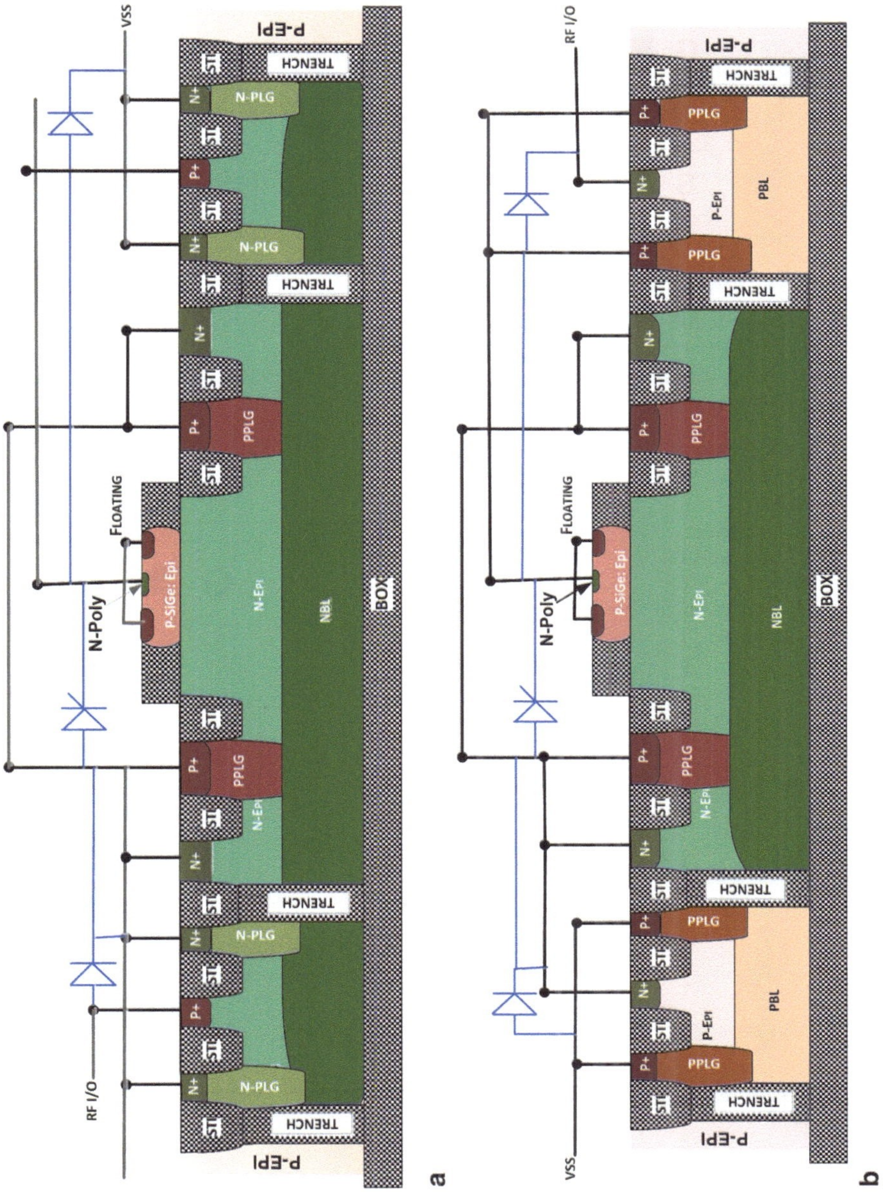

Fig. 3.13 SiGe SCR with blocking diodes for microwave I/Os. **a** Two forward p-diodes in series with SCR for protection between RF I/Oand VSS. **b** Two forward n-diodes in series with SCR for protection from VSS to RF I/O

High frequency IOs with operating signal swinging around a particular DC point require other considerations for the ESD protection. Since the signal swing requirements are also stringent at high temperature, the diode stack is less optimum. The Included bidirectional SCRs formations or SCRs in antiparallel configuration

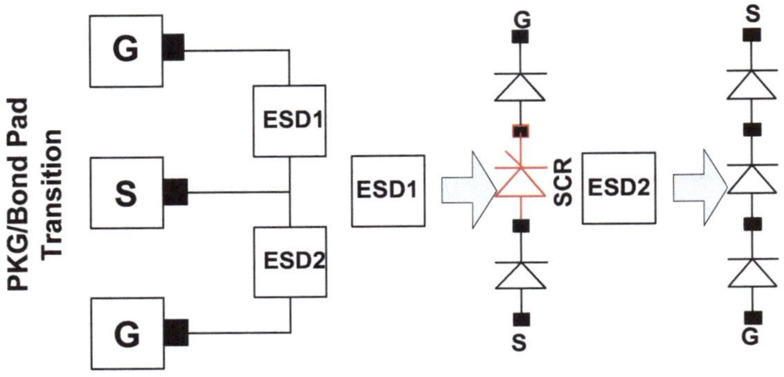

Fig. 3.14 ESD protection in ground-signal-ground (*GSG*) configuration for large signal (>+3 V) microwave I/O applications

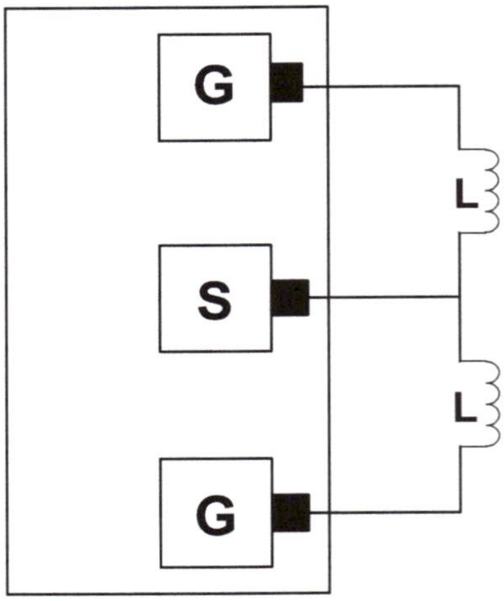

Fig. 3.15 ESD protection for ±1 V microwave I/Os (*narrow band*). The value of L depends on the frequency window

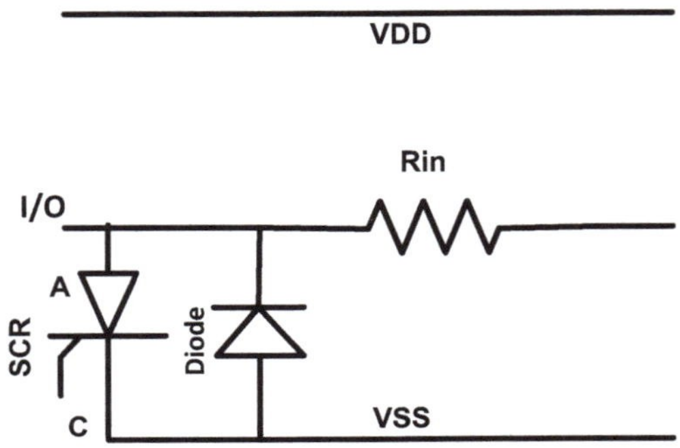

Fig. 3.16 SCR and diode I/O protection

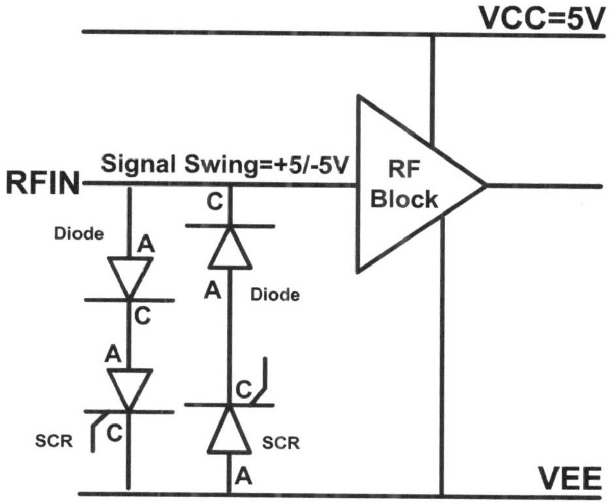

Fig. 3.17 +5/−5 V RF I/O protection

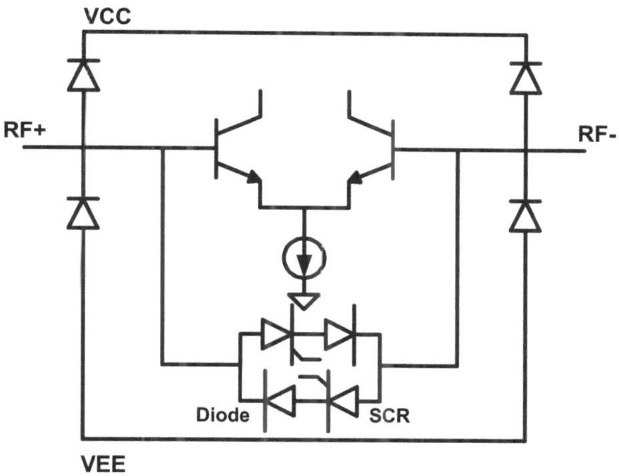

Fig. 3.18 2.5 V differential RF I/O protection

[36] is a better option in these cases. Figures 3.17 and 3.18 show the SCR devices arranged in various configurations to protect large signal swing RF IOs, for example, ±5 input signal swing and 2.5 V differential RF signal. In these cases, the SCR devices provide the higher breakdown voltage over temperature. In the case of Figs. 3.17 and 3.18, a series ESD blocking diode is included for allowing relatively high breakdown voltage for bidirectional signal when the diode-SCR combination are arranged in antiparallel configuration.

3.4 Summary

In this chapter, SiGe-based SCRs (SiGe SCRs) fabricated in a 0.18-μm BiCMOS process were analyzed using the vfTLP and TCAD simulation. It was demonstrated that a SiGe SCR implemented with a P PLUG layer and minimal PNP base width can offer the smallest peak voltage and fastest response time. The improved performance mainly resulted from the fact that the impact ionization region and effective base width in the SiGe SCR were reduced due to the presence of the P PLUG layer. A practical approach for integrating optimum ESD protection solutions for the low-voltage/high-speed SiGe-based integrated circuits was subsequently illustrated using conventional RF circuit pads configurations and circuit blocks.

Chapter 4
On-Chip Radio Frequency ESD Protection Solution in GaAs pHEMT Process

4.1 Failure Analysis of Conventional Single-Gate pHEMT-Based ESD Protection Clamp

Electrostatic discharge (ESD)-induced failure is a major concern for the pseudomorphic high electron mobility transistor (pHEMT)-based integrated circuits [60, 61]. This reliability issue is further worsened by the inherent low thermal conductivity of compound materials like gallium arsenides (GaAs) and aluminum gallium arsenides (AlGaAs). Nowadays, there has been a growing demand for the availability of robust ESD protection solutions for radio frequency integrated circuits (RFICs) fabricated in the pHEMT process. Stacked Schottky diodes could provide ESD protection in the GaAs pHEMT process. However, this approach has the disadvantages of a huge layout area, high leakage current, and large on-state resistance. On the other hand, a structure consisting of a shunting apparatus and external trigger unit is a more effective ESD protection scheme (see Fig. 4.1). A conventional single-gate GaAs pHEMT is usually used as the shunting apparatus [62], but the current conducting capability of such a device, and hence the GaAs pHEMT-based ESD protection clamp, is still quite limited due to the self-heating and current saturation mechanisms [63].

4.1.1 *pHEMT Clamp Technology Computer-Aided Design (TCAD) Setup for Failure Analysis*

Technology computer-aided design (TCAD) simulation tools have been widely used in assisting the development of devices and circuits for over three decades. Typically, a TCAD tool consists of process and device simulators. The process simulator generates the device structure based on given fabrication steps (e.g., lithography, ion implantation, temperature). The simulated device structure is input to the device simulator, which then generates the meshes and solves a set of fundamental

Q. Cui et al., *On-Chip Electro-Static Discharge (ESD) Protection For Radio-Frequency Integrated Circuits*, DOI 10.1007/978-3-319-10819-3_4

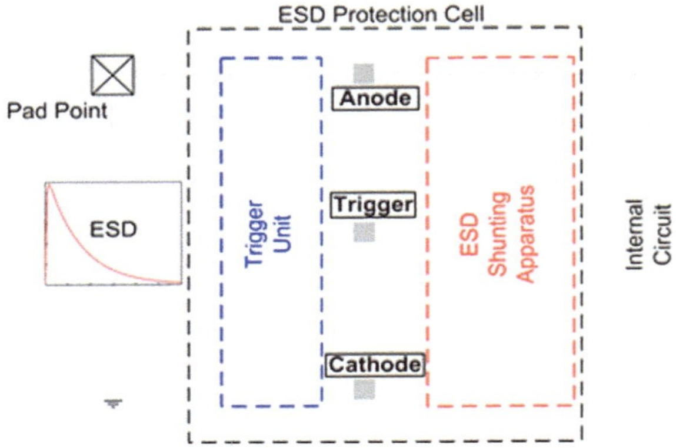

Fig. 4.1 ESD protection structure with trigger unit in pHEMT process

semiconductor equations for every mesh point. These equations include the Poisson and continuity equations. For ESD applications, a TCAD software tool is more advantageous over a circuit-level simulator (e.g., SPICE). This is because ESD devices operate under a high-current and high-voltage condition, and the compact models available in SPICE are typically unable to support such an operation. Since the TCAD simulation is more physics-based, it is theoretically not limited to any biasing conditions. In this book, the Sentaurus software tool supported by Synopsis will be employed for TCAD simulation.

As mentioned in Chap. 1, the transmission line pulsing (TLP) tester is an effective tool to characterize the device behaviors under an ESD event. Here we will analyze the single-gate GaAs pHEMT-based ESD protection clamp using both TLP measurement and TLP-like TCAD simulation.

Figure 4.2a shows the setup for measuring the GaAs pHEMT using the Barth 4002 TLP tester, which generates human body model-like pulses with a 100-ns pulse width and 10-ns rise time. Figure 4.2b, c illustrates the direct current (DC)-like and TLP-like TCAD simulation setups, respectively. The 50-Ω resistor Rd is used to imitate the 50-Ω characteristic impedance associated with the transmission line of the TLP system. In the DC-like simulation, ramped voltages are directly applied to terminals of ESD devices, and the lattice temperature module is turned off. This simulation takes less time and is accurate enough when the device is operating under a low-power condition. However, the DC-like simulation has two inherent disadvantages:

i. DC-like simulation cannot represent the time-dependent characteristics correctly. For instance, the DC-like simulation can never reflect the dynamic behavior of the device under test (DUT) when it is subject to TLP pulses.
ii. DC-like simulation is difficult to account for the thermal effect, as the simulation of temperatures in a device is strongly related to the transient response of heat generation and dissipation.

As a result, TLP-like simulation will be carried out here to overcome the aforementioned problems. Figure 4.3 shows schematically the TLP pulses applied to the

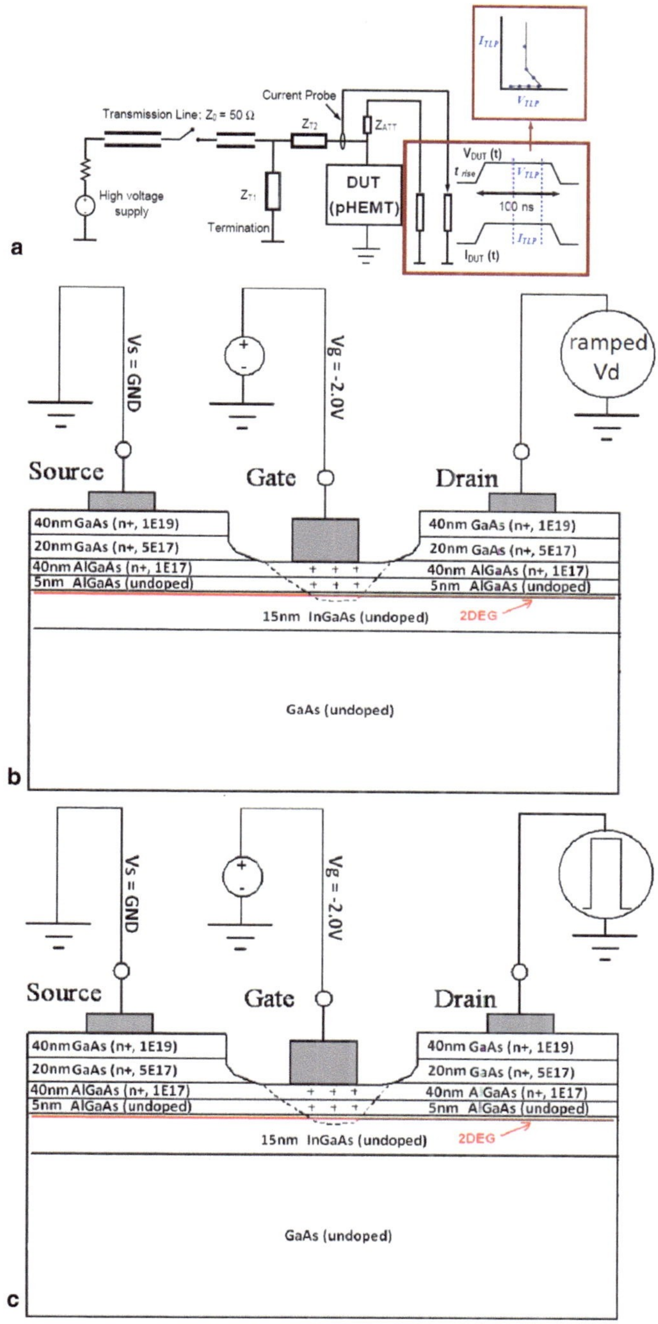

Fig. 4.2 **a** Measurement setup using Barth 4002 TLP. **b** DC-like simulation setup. **c** TLP-like simulation setup

Fig. 4.3 TLP pulses used, time-dependent voltage and current waveforms simulated, and quasistatic current-voltage curve extracted from the voltage and current waveforms

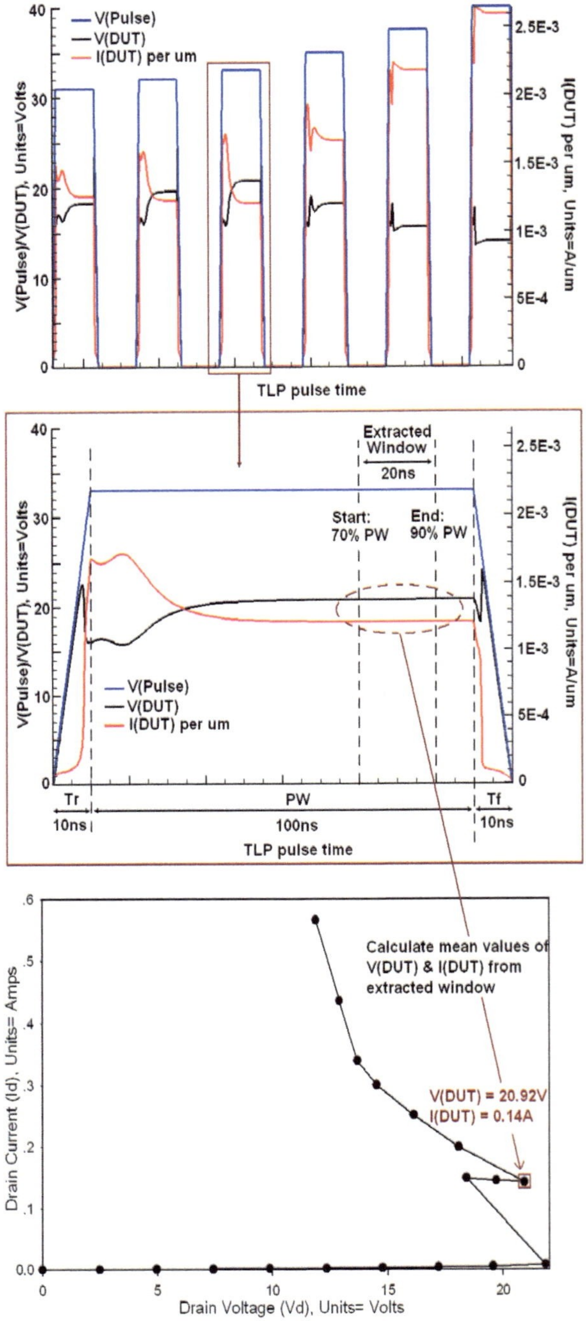

device under test, the simulated time-dependent voltage and current waveforms, and the quasistatic current vs. voltage curve extracted from the voltage/current waveforms. Note that the TLP current–voltage (I–V) data are extracted from the voltage/current vs. time waveforms in the area where the waveforms are fairly insensitive to time and by averaging the voltages and currents in this window. Beyond the trigger point, the voltage decreases. This phenomenon is called the snapback, which is also observed in TLP measurements.

Figure 4.4a compares the simulated and measured quasistatic I–V curves of the GaAs pHEMT under two different gate bias conditions. The self-heating effect has been included in the TLP-like simulation. Good agreement is found between simulation and measurement results. Snapback (points A to B) and postsnapback saturation (points B to C) behaviors are clearly illustrated. It should be noted that the device failed beyond point C. In contrast, the DC-like simulation given in Fig. 4.4b predicts no postsnapback saturation due to its inability to incorporate temperature effect. Figure 4.4b further demonstrates that without including the temperature effect, the TLP-like simulation is quite accurate in the presnapback region but highly questionable in the postsnapback operation because of the existence of very large current in such an operation.

4.1.2 *Snapback Behavior*

Snapback is an I–V phenomenon in which the device's voltage decreases as the device current increases. Snapback is widely seen in ESD devices. For example, the snapback found in a metal–oxide–semiconductor (MOS) device can be attributed to the ESD-induced triggering of the parasitic bipolar junction transistor (BJT). Similarly, snapback behavior is also found in the pHEMT under study (see Fig. 4.5). The BJT-triggered snapback mechanism in a MOS device cannot be directly applied to the pHEMT because of the absence of a parasitic BJT imbedded in the pHEMT. However, we will prove that there is a virtual parasitic BJT in the pHEMT under high ESD stress. Figure 4.5a and b shows the pHEMT cross-sectional view and equivalent BJT-like structure formed along the ESD current discharging path, respectively. The base region of the virtual BJT is formed by the holes generated by heavy impact ionization near the drain region.

When the pHEMT is subject to a TLP pulse, the holes and electrons generated by the impact ionization are driven toward the source and drain, respectively, as a result of the lateral electric field due to the elevated drain voltage. Figure 4.6a includes the schematic of the 0.5-μm InGaAs depletion mode pHEMT under study, including the bias connections used in the simulation and simulated temperature contours at the snapback point "A." Figure 4.6b shows that the simulated hole density near the AlGaAs/InGaAs interface under the gate increases significantly in the postsnapback regime (from points A to B in Fig. 4.4). This in turn forms the base region for the virtual parasitic BJT. As the hole density in this base region increases, the potential in the InGaAs layer rises. This lowers the potential barrier to the electrons in the AlGaAs layer, as demonstrated in Fig. 4.6b using the simulated conduction band

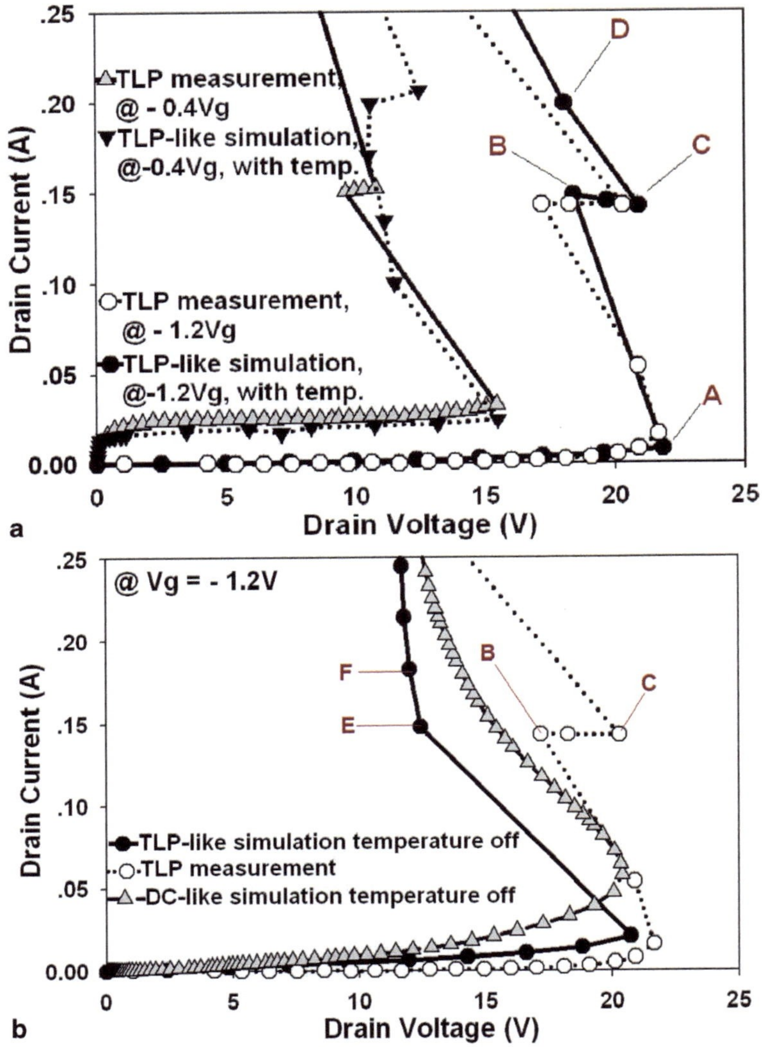

Fig. 4.4 a Comparison of the measured and TLP-like simulated I–V curves, with the snapback (points *A* to *B*) and postsnapback saturation (points *B* to *C*) indicated. In both measurement and TLP-like simulation, square pulses with a 10 ns rise time and 100 ns pulse width were applied. **b** Comparison of pHEMT's I–V curves obtained from TLP measurement, DC-like simulation, and TLP-like simulation without including the electrothermal simulation module

energies at the source side before and after snapback. This allows for a high-level injection of holes from the InGaAs base to the AlGaAs emitter, and a large number of electrons are injected in the opposite direction. The conduction of such a virtual BJT reduces the voltage drop between the drain and source terminals, resulting in the observed snapback behavior.

To provide more detailed insights, Fig. 4.7a, b shows the simulated impact ionization and electric field contours, respectively, at the onset snapback point

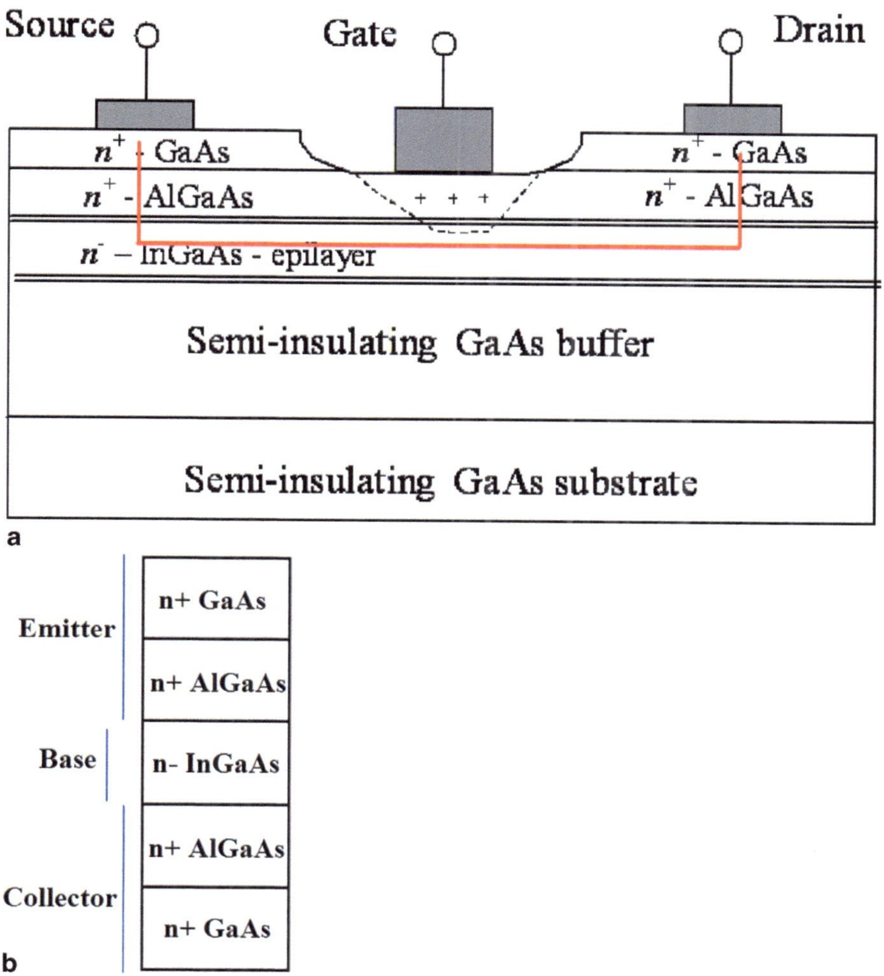

Fig. 4.5 **a** Cross section of pHEMT and its ESD current discharging path, and **b** parasitic BJT formed along the current path

(point A in Fig. 4.4) and postsnapback point (point B in Fig. 4.4). Both the high-impact-ionization and high-electric-field regions move away from the virtual BJT's base/collector junction toward the collector region as the current increases. This further verifies the virtual bipolar effect in the pHEMT operating under the ESD stress. This is analogous to the Kirk effect phenomenon in typical Si bipolar devices.

4.1.3 Postsnapback Saturation Behavior

Beyond snapback, there is a unique current saturation behavior observed prior to failure (see Fig. 4.8a). This region is named the postsnapback saturation region in

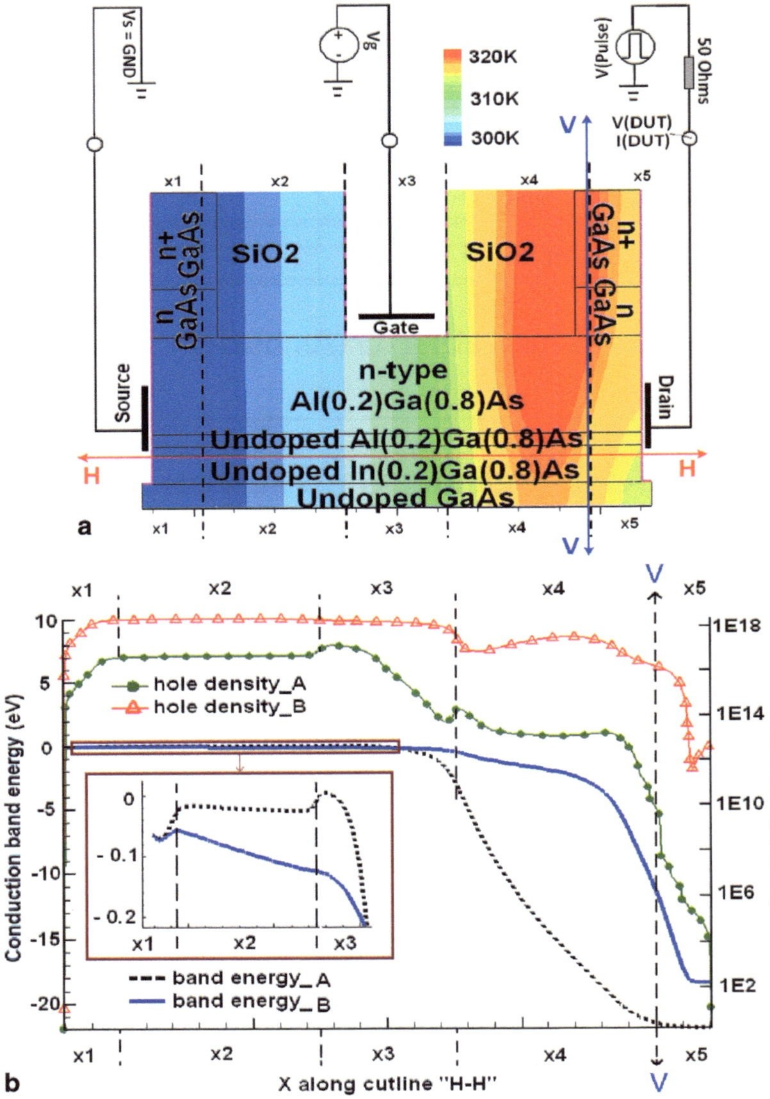

Fig. 4.6 **a** Schematic of the 0.5-µm InGaAs depletion mode pHEMT under study, including the bias connections used in the simulation and simulated temperature contours at the snapback point A. **b** Simulated hole densities and conduction band energies along the horizontal line (*H-H*) at onset of snapback ("A") and postsnapback ("B"). The *insert* is the zoom-in of the band energies in regions *x1*, *x2*, and *x3*

order to differentiate it from the traditional saturation region. It will be proved that the postsnapback saturation is caused by the temperature-induced velocity saturation. Diagram flow for this hypothesis is shown in Fig. 4.8b. As well known, the electron current density is proportional to the product of electron mobility, electric

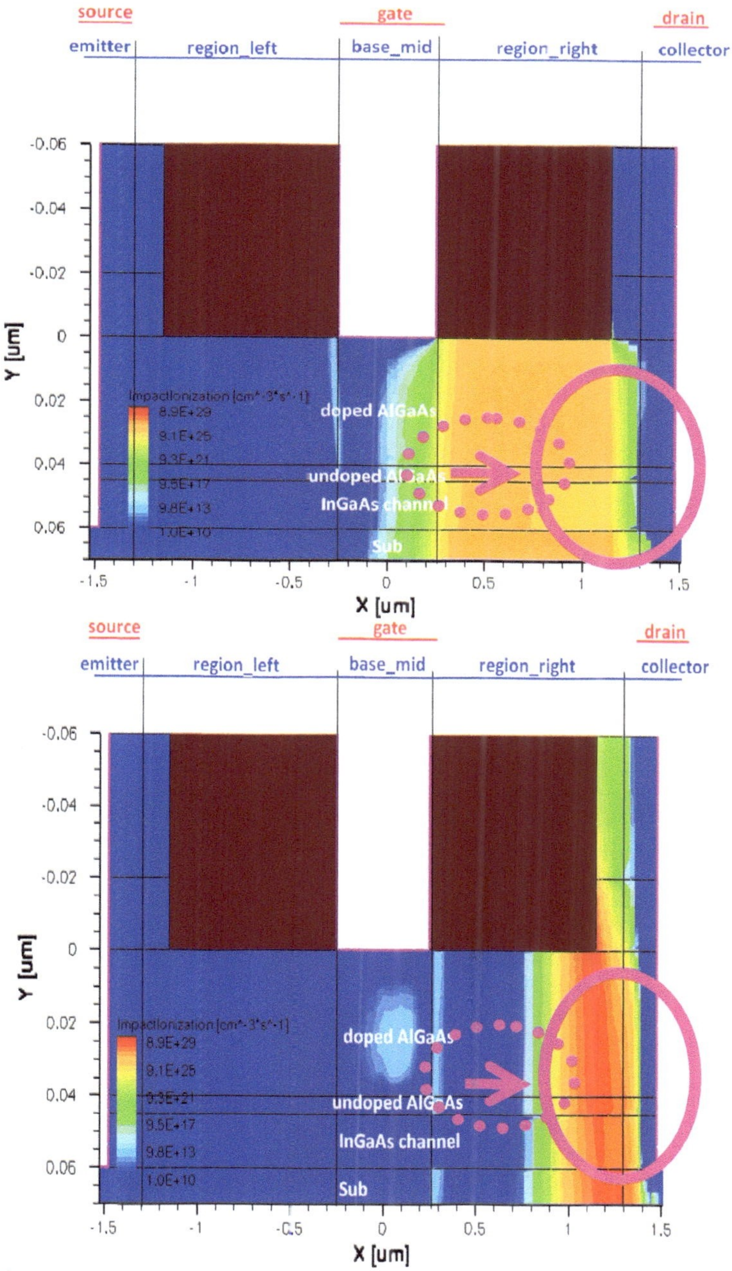

a

Fig. 4.7 **a** Simulated impact ionization contours at point "A" (*top*) and at point "B" (*bottom*). **b** Electric field contours at point "A" (*top*) and at point "B" (*bottom*)

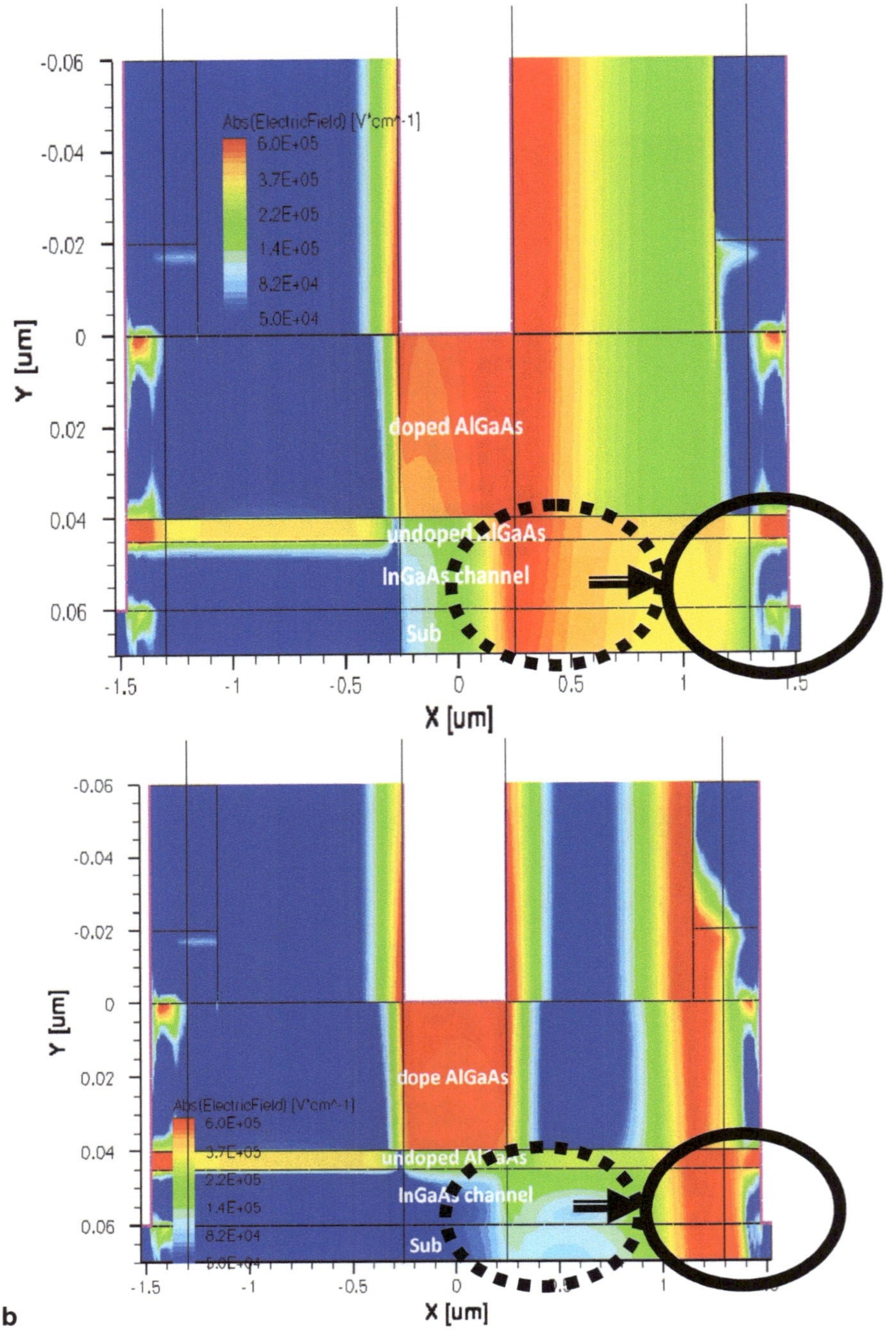

Fig. 4.7 (continued)

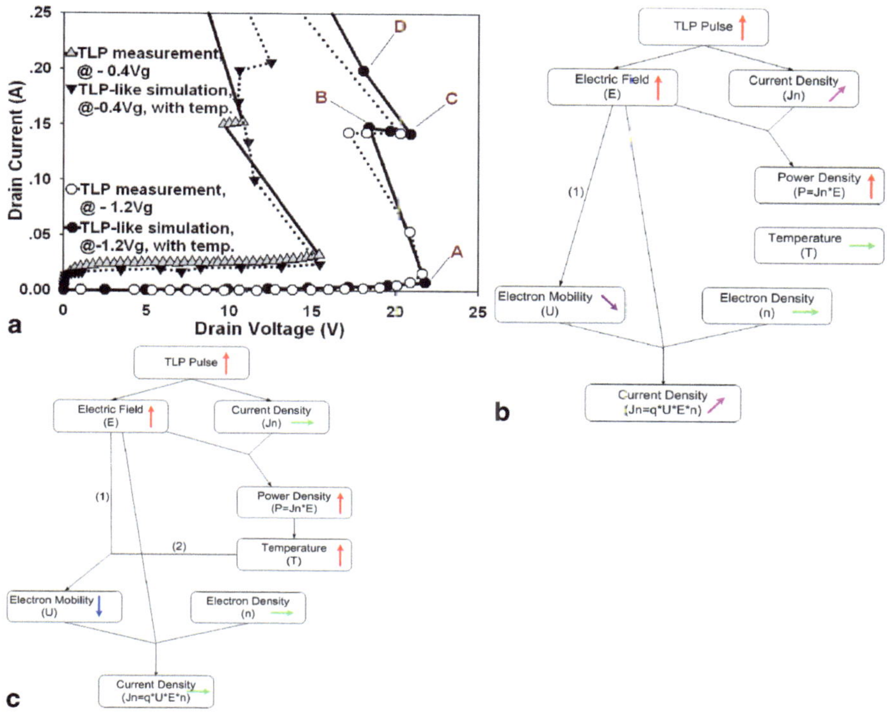

Fig. 4.8 Postsnapback saturation mechanism of 0.5 μm depletion mode pHEMT. **a** Comparison of the measured and TLP-like simulated I–V curves. **b** Explanation of TCAD simulation with lattice temperature module "OFF." **c** Explanation of TCAD simulation with lattice temperature module "ON"

field, and electron density (Jn=q*μ*E*n). In Fig. 4.8b, the lattice temperature module is turned off. Therefore, increasing power density cannot raise the lattice temperature under TCAD simulation. We select two points (points E and F) from the simulated I–V curve without the temperature effect in Fig. 4.4b. The relevant physical mechanisms involved from points E to F are:

i. Electric field in the channel near the drain region increases considerably. This is because increasing TLP-like voltage raises the pHEMT's drain voltage Vd.
ii. Electron mobility decreases as the electric field increases, according to the semiconductor theory.
iii. Electron density is nearly the same because of the fact that almost all the electrons have been depleted.
iv. Since the electron current density equals the product of electric field, electron mobility, and electron density, we cannot see the postsnapback saturation when the temperature module is turned off.

In Fig. 4.8c, the lattice temperature module is turned on. We again select two points (points B and C) from the simulated I–V curve in Fig. 4.4a. Physical mechanisms involved from points B to C are:

i. Electric field increases sharply due to the increased pHEMT's drain voltage Vd.
ii. Electron mobility decreases considerably because of the following two reasons: increased electric field and increased lattice temperature.
iii. Electron density is nearly the same because of the fact that almost all the electrons have been depleted.
iv. Since the electron current density equals the product electric field, electron mobility, and electron density, we can anticipate postsnapback saturation when the temperature module is turned on.

Figure 4.9a, b shows the electric field and lattice temperature contours, respectively, in the postsnapback saturation region. In the region marked with *V–V,* both the electric field and lattice temperature are high, and the electron velocity is saturated. The simulated averaged electric field (E), electron mobility (μ), electron density (n) and electron current density (Jn) along the cutline V–V can be obtained. The trends of electron mobility vs. temperature and electric field are plotted in Fig. 4.10a and b, respectively. Here, the electron mobility is found to decrease with the increase of temperature and electric field.

Table 4.1 lists the simulated parameters of pHEMT, which were obtained by averaging the data along the V-V line near the drain junction operating at points A, B, C, and D. Note that prior to the virtual BJT formation, at point A, the drift current accounts for over 90 % of the total current, which indicates a future and emerging technologies (FET) action. After the virtual BJT is turned on at point B, the diffusion current accounts for over 75 % of total current, which indicates the existence of an BJT action. From points B to C, the electron and hole densities remain constant, the electric field continues to increase with the increase in drain voltage, but this is offset by the decrease in the mobility due to increasing temperature (Table 4.1). Because of this electrical–thermal interacting effect, the diffusion current also saturates. This is the result of electron and hole densities being constant, as the effects of temperature increase and mobility decrease offset each other. Saturation of both the drift and diffusion currents makes the total current constant.

4.1.4 *Thermal Failure Behavior*

From points C to D in Fig. 4.4, the decrease of voltage drop and the increase of current are due to the significant conductivity modulation in the pHEMT. At the same time, the device has entered the intrinsic generation region due to very high temperature, and both the electron and hole densities have increased to an uncontrollable level. This unstable operation leads to a catastrophic hot spot and device failure. In TLP measurements, there is no point D, as the pHEMT fails immediately after point C. We will explain the thermal-induced failure behavior using TCAD simulation and scanning electron microscope (SEM) analysis.

pHEMT has multiple layers, and each layer has a different melting point (see Table 4.2). In order to locate the hot spot in pHEMT, we set the failure criteria in

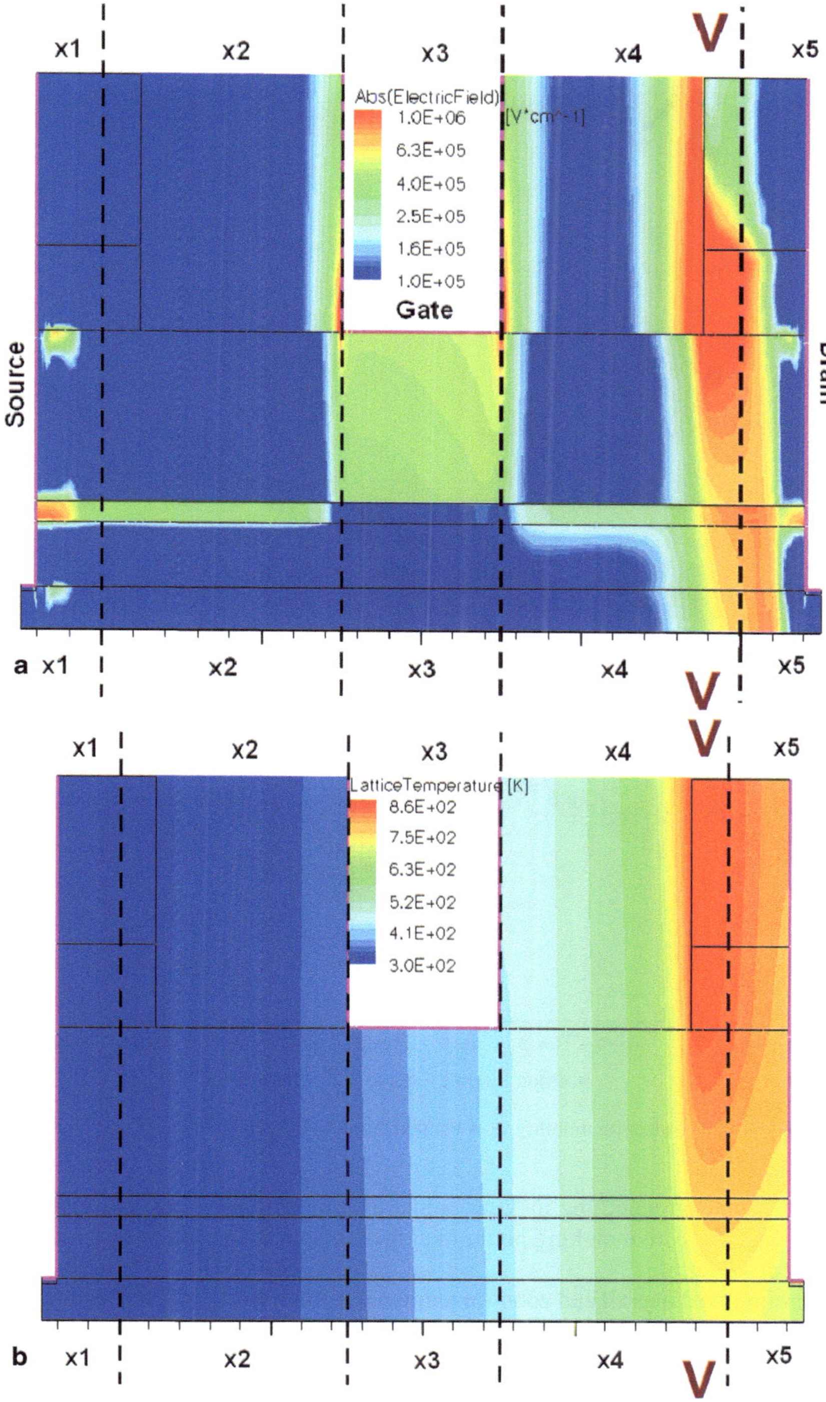

Fig. 4.9 TCAD Simulated **a** electric field **b** temperature contours in the post-snapback saturation region

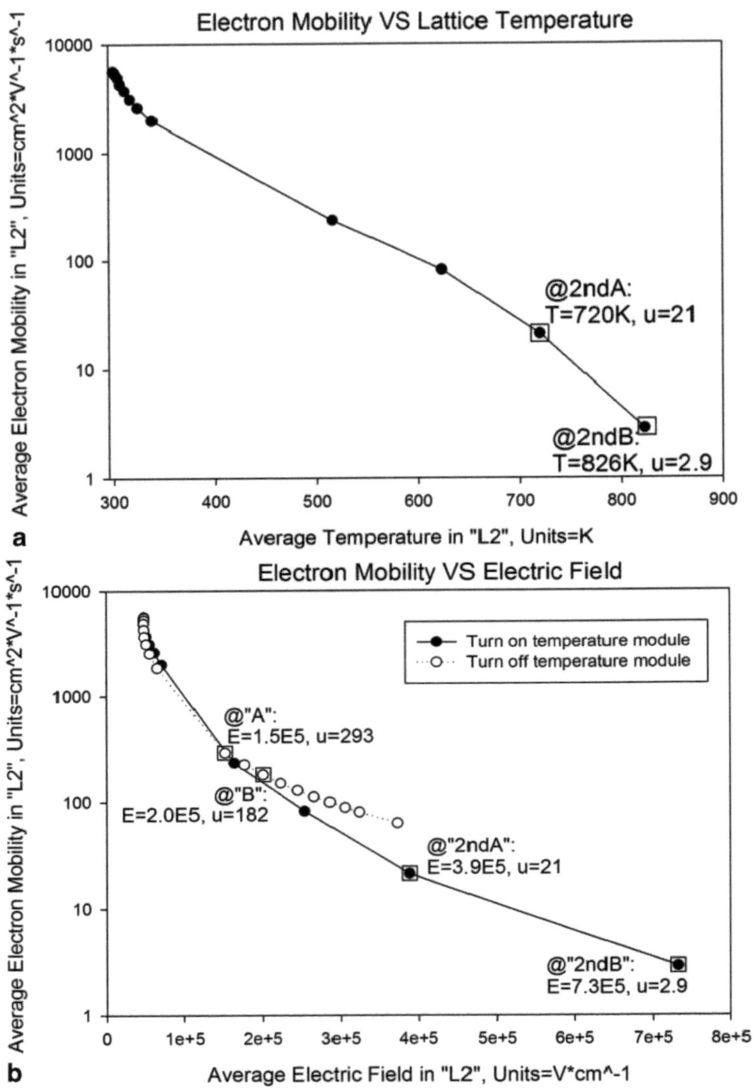

Fig. 4.10 Simulated electron mobility vs. **a** temperature and **b** electric field

TCAD simulation using the melting points. If temperature in any given material exceeds the respective melting point, then the hot spot is located and the device is damaged.

The simulated overall and zoom-in temperature contours of pHEMT operating at point D are plotted in Fig. 4.11a and b respectively, and the hot spot image obtained from SEM is given in Fig. 4.11c. The simulation and failure analysis results correlate well with each other. We can see that the hot spots are located at the edge of

Table 4.1 Simulated pHEMT parameters at points A, B, C, and D, where |E| is the electric field, Jn and Jp are the electron and hole current densities, n and p are the electron and hole densities, and μn and μp are the electron and hole mobilities

Parameter	A	B	C	D
Temp (K)	317	787	826	908
|E| (V/cm)	7.2E4	5.3E5	8.0E5	7.8E5
Jn (A/cm2)	4.23E4	5.92E5	5.78E5	6.76E5
Jn_diffu (A/cm2)	2.82E3	5.36E5	5.36E5	4.79E5
Jn_drift (A/cm2)	3.95E4	4.23E4	4.23E4	1.94E5
μn (cm2/(V s))	2.7E3	8.33	2.38	1.89
n (cm -3)	3.53E18	3.22E18	3.27E18	4.03E18
Jp (A/cm2)	1.07E3	1.13E5	1.13E5	2.28E5
Jp_diffu (A/cm2)	56.4	2.82E3	5.64E3	1.16E5
Jp_drift (A/cm2)	1.01E3	1.10E5	1.07E5	1.12E5
μp (cm2/(V s))	2.0E3	41.0	36.1	31.6
p (cm −3)	3.64E8	1.13E16	1.14E16	2.38E16
Jtotal (A/cm2)	4.34E4	7.05E5	6.91E5	9.04E5

Table 4.2 Melting temperatures in pHEMT

Materials in pHEMT	Melting point (K)
GaAs	1511
SiO2	1873
Si3N4	2173
AlGaAs	1523
InGaAs	1373
Gold (Drain metal)	1336

the drain metal (gold). The simulated temperature at the gold contact exceeds gold's melting point 1336 K. Since the melting points of the gold and InGaAs layer are almost the same, damage can happen in both materials.

4.2 Multigate E-Mode pHEMT-Based ESD Protection Clamp

This section reports a new multigate GaAs pHEMT, which can be used as an effective device in protecting GaAs-based ICs against ESD threats. With approximately the same layout area and parasitic capacitance, the proposed ESD protection device can carry a current three times higher than the conventional single-gate pHEMT device. Moreover, the new device possesses a very low on-state resistance, fast turn-on speed, low parasitic capacitance normalized to failure current, and flex-

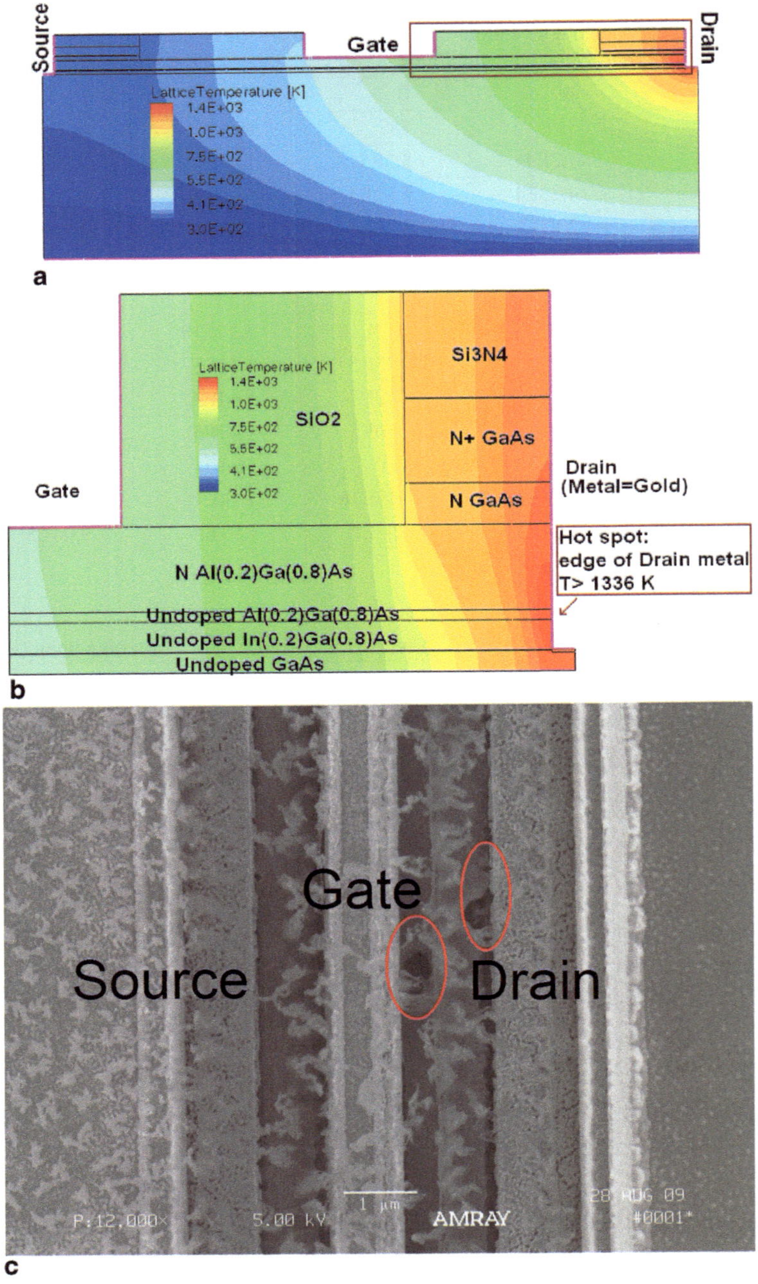

Fig. 4.11 **a** Simulated overall pHEMT temperature contours. **b** Simulated zoom-in temperature contour near the drain region. **c** Hot spots obtained using SEM

ibility to adjust the trigger voltage for fulfilling different ESD protection design requirements.

4.2.1 Device Structure

Figure 4.12a, b shows the cross-sectional view of a conventional enhancement-mode, single-gate, GaAs pHEMT and schematic of an ESD clamp built based on such a device, respectively. The ESD clamp consists of a diode chain and a resistor (i.e., current limiter) to control the triggering of the pHEMT. Under normal operating conditions, the diode chain (see Fig 4.12b) is not turned on, potential at the trigger terminal is low, pHEMT is off, and current flow between the anode and cathode is negligibly small. Under the ESD condition, a stress voltage at the anode

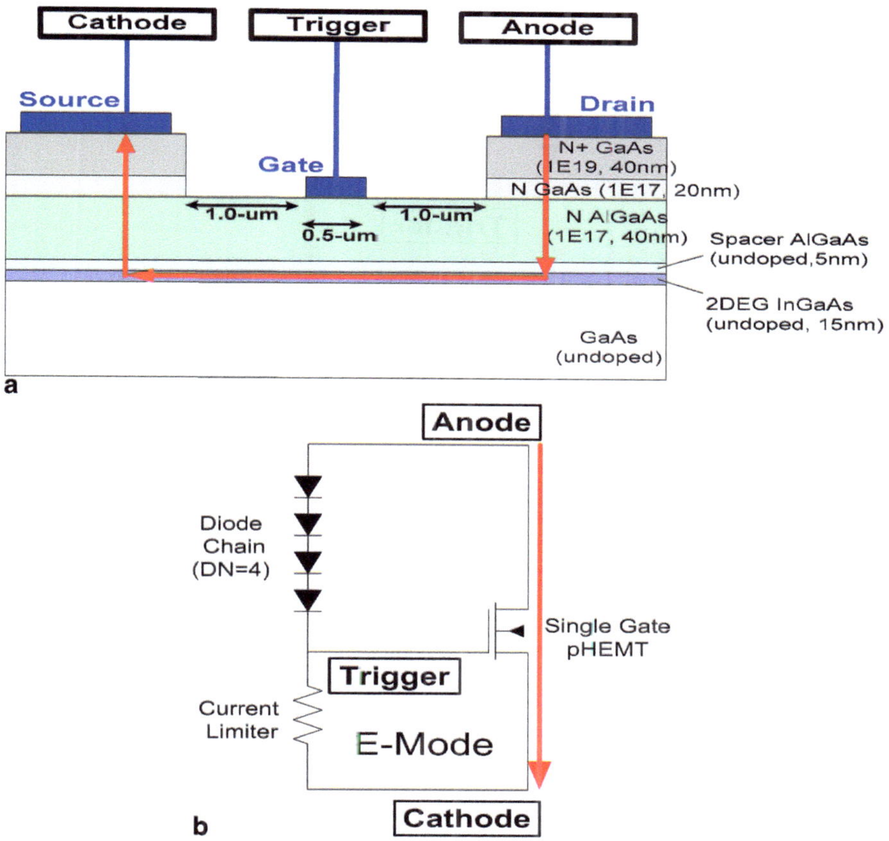

Fig. 4.12 **a** Cross-sectional view of single-gate pHEMT with the *red line* denoting the current path. **b** ESD clamp built using the single-gate pHEMT and diode chain

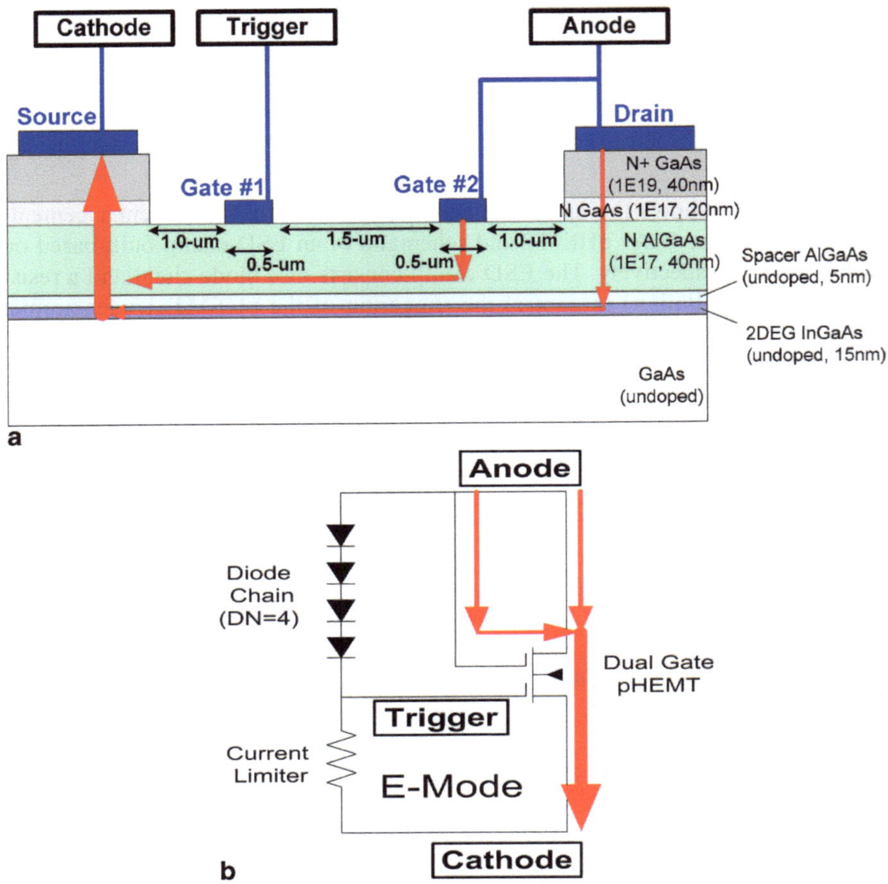

Fig. 4.13 **a** Cross-sectional view of proposed dual-gate pHEMT with the two *red lines* denoting the two current paths. **b** ESD clamp built using the dual-gate pHEMT and diode chain

larger than the blocking voltage of the diode string first turns on the diodes and then forces a current through the resistor. This raises the potential at the trigger terminal, reduces the depletion region underneath the gate, exposes the 2DEG channel, and consequently allows for ESD current to discharge between the anode and cathode via the 2DEG channel (indicated by the red line in Fig. 4.12). Such an ESD clamp is considered a better protection design than the Schottky diode-string-based clamp [62, 64], but its robustness is still relatively poor due to the fact that there is only one current discharging path (via the 2DEG channel) in the single-gate pHEMT.

Figure 4.13a, b shows the cross-sectional view of the proposed enhancement-mode, dual-gate, GaAs pHEMT and the schematic of its ESD clamp, respectively. Gate #1 in the dual-gate pHEMT is the same as the gate in the single-gate pHEMT. The new feature in this device is a second gate, Gate #2, connected to the anode (Fig. 4.13a). A similar concept was reported by Lin et al. [65] aiming at reducing the pHEMT's parasitic capacitance. The work presented here will show that the intrinsic advantages of such a device are actually the relatively high failure current and low on-state resistance, rather than the small capacitance. Under a sufficiently large ESD stress voltage, the Schottky junction underneath Gate #2 will conduct. This forms a second current discharge path from Gate #2 through the AlGaAs layer to the source (see Fig. 4.13b), in addition to the 2DEG current discharging path created by Gate #1. The added current path reduces the on-state resistance and increases the robustness of the dual-gate ESD clamp. More detailed explanations for the mechanisms of two current conducting paths will be given in the next section.

4.2.2 Measurements and Discussion

Single- and dual-gate (SG and DG) pHEMT clamps were fabricated in a 0.5-μm pHEMT process at WIN Semiconductor. A triggering unit consisting of four-diode chain and a current limiter was incorporated in both clamps. The diode area was 62×53 μm², the current limiter resistor area was 40×20 μm², and the single- and dual-gate pHEMT each had 20 fingers and areas of 116×106 μm² and 135×106 μm², respectively. For the GaAs technology used, the smallest size available for the dual-gate device was 24×3 μm².

ESD measurements were carried out using the Barth 4002 transmission line pulsing (TLP) tester which generates human body model (HBM)-like pulses having a 10 ns rise time and 100 ns width. Poststress leakage currents were measured at a voltage of 2.2 V. This voltage is the pHEMT normal operating voltage of 2.0 V, plus a 10% dynamic voltage margin. In addition, the Hanwa HED-W5000M on-wafer HBM tester was used to accurately assess the ESD clamps' robustness.

Figure 4.14a compares the TLP current density (current per width, *y*-axis) vs. voltage (bottom *x*-axis) and leakage current (top *x*-axis) curves of the single- and dual-gate ESD clamps. Clearly, the dual-gate clamp possesses a smaller on-state resistance and higher failure current than its single-gate counterpart. The failure current densities (failure current I_{t2} per width) of the single- and dual-gate devices are 1.15 mA/μm and 3.8 mA/μm, respectively. The insert in Fig. 4.14a shows the TLP I–V curves near the triggering point. Note that the trigger voltage depends on the number of diodes in the diode chain, and this voltage can be adjusted by altering the diode number and diode polarity. The trigger voltage of about 3.2 V observed in Fig. 4.14a resulted from the use of a four-diode string. It is noticed that the resistances of both single-gate and dual-gate clamps change at around 5 V. This can be attributed to a positive feedback due to the electron-hole pair generation near the drain junction and the turn-on of embedded parasitic lateral BJT [63]. Figure 4.14b shows the TLP I–V characteristics of DG pHEMT's having four different diode strings. While the trigger voltage depends on the diode number, the failure current is insensitive to such a factor. Table 4.3 summarizes the measured trigger voltage as a function of diode number and polarity.

Of another important phenomenon found in Fig. 4.14 is that the I–V curves of single- and dual-gate clamps have almost the same slopes when the voltage is less

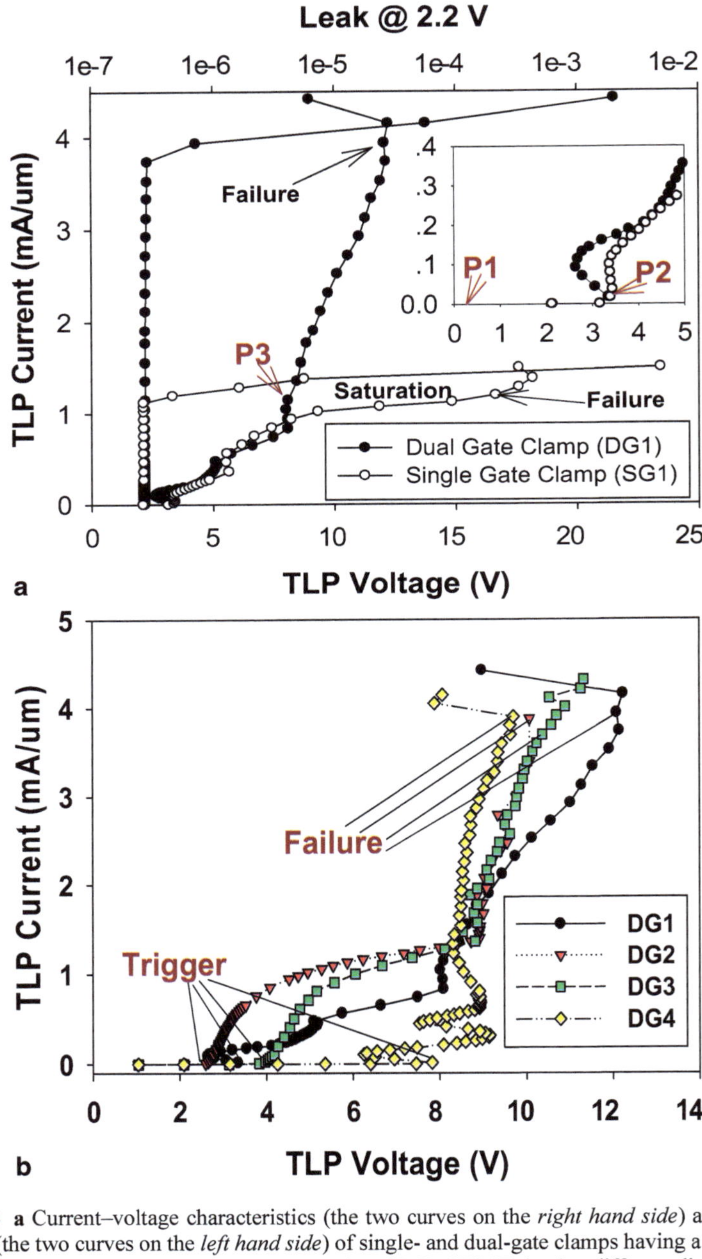

Fig. 4.14 **a** Current–voltage characteristics (the two curves on the *right hand side*) and leakage currents (the two curves on the *left hand side*) of single- and dual-gate clamps having a four-diode string obtained from the TLP tester. **b** TLP results of DG pHEMTs having different diode number and polarity (see DG1, DG2, DG3, DG4 in Table 4.3)

Table 4.3 Trigger voltage dependency on diode connection for E-mode pHEMT ESD clamp

Dual-gate clamp	Diode polarity	Diode number	Trigger voltage of single-gate pHEMT (V)	trigger voltage of dual-gate pHEMT (V)
DG1	Forward	4	3.2	3.2
DG2	Forward	3	–	2.5
DG3	Forward	5	–	4.1
DG4	Reverse	1	7.5	7.5

than 8 V. Beyond this, the single gate clamp enters the saturation region caused by temperature induced mobility degradation [63], whereas the dual-gate clamp exhibits a steeper curve stemming from the turn-on of the second current path.

The different current conduction mechanisms can be explained in more details below using physical reasoning and TCAD simulation results. When the stress voltage is equal to the threshold voltage of pHEMT, 0.25 V, the 2DEG channel underneath Gate #2 is unblocked but underneath Gate #1 is still obscured by the depletion region because of the blockade of four-diode chain (see schematic in Fig. 4.15a and simulated electron density in Fig. 4.16a). The Schottky diode of Gate #2 is also off at this time. When the stress voltage is increased to 3.2 V, the clamp trigger voltage, the Schottky diode begins to turn on but its current path from the Schottky contact to the source via AlGaAs layer is still blocked by the depletion region associated with Gate #1 (see schematic in Fig. 4.15b and simulated electron density in Fig. 4.16b). The 2DEG underneath Gate #1 is unveiled at this time because the diode chain is conducting. So there is one current conducting path between the anode and cathode via the 2DEG channel (see Figs. 4.15b and 4.16b). As the stress voltage at the anode reaches 8 V, where the slope of I–V curve of DG1 is increased sharply, the depletion region associated with Gate #1 also vanishes, and the second current path associated with the Schottky diode is formed (see Figs. 4.15c and 4.16c). As a result, two current conducting paths exist in the dual-gate pHEMT when the stress voltage is larger than 8 V, giving rise to a smaller on-state resistance and larger failure current.

The robustness of the two ESD clamps was further verified using the Hanwa HED-W5000M on-wafer HBM tester. The single-gate clamp can survive up to a 2 kV HBM zap, whereas the new dual-gate clamp can pass 8 kV HBM zap. These results are consistent with the trend found in Fig. 4.14, but they are not exactly equal to the HBM passing voltages obtained by the conventional estimate of multiplying the TLP failure currents to 1500 Ω human body resistance due to the different setup and calibration issues.

Figure 4.17 presents the voltage/current waveforms of the two clamps zapped at 1 kV obtained from the on-wafer HBM tester. The maximum clamping voltage of single-gate clamp SG1 is 22 V, which is similar to data reported in [62] for a single-gate pHEMT fabricated in a comparable GaAs technology. The lower maximum clamping voltage of 12 V of the dual-gate clamp DG1 reduces the likelihood of damages to the core circuit. The reason SG1 has a much higher clamping voltage than DG1 is because SG1 enters the saturation at 8 V while at the same voltage DG1 continues its linear I–V characteristics (see Fig. 4.14), thus giving rise to a small voltage drop in DG1.

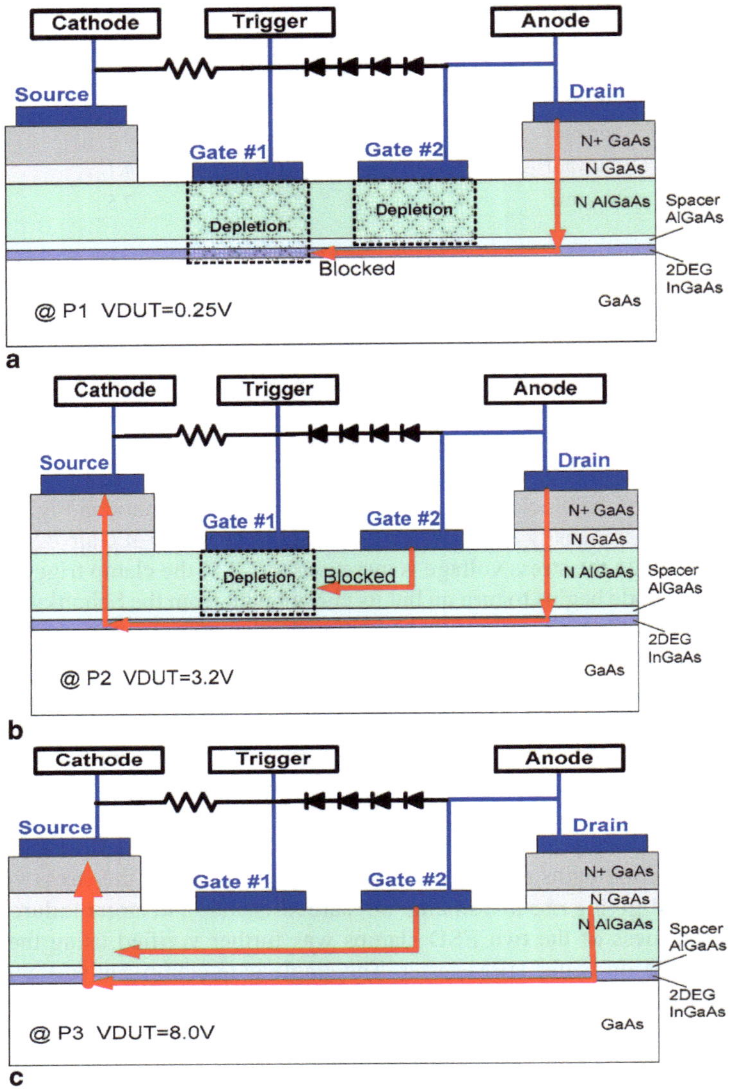

Fig. 4.15 Depletion regions and current paths in the dual-gate pHEMT at a stress voltage of 0.25 V (**a**), 3.2 V (**b**), and (**c**) 8.0 V (**c**)

Another interesting characteristic found in Fig. 4.17a is that the voltage vs. time waveform of the single-gate clamp SG1 suddenly drops at 55 ns while the dual-gate clamp DG1 does not. This observation can be explained below. As shown in Fig. 4.14a, SG1 enters the current saturation region when the current is higher than about 0.5 mA/μm (i.e., equivalent to 0.5 A). The saturation results from the flow of discharging current being limiting to the 2DEG channel and mobility deg-

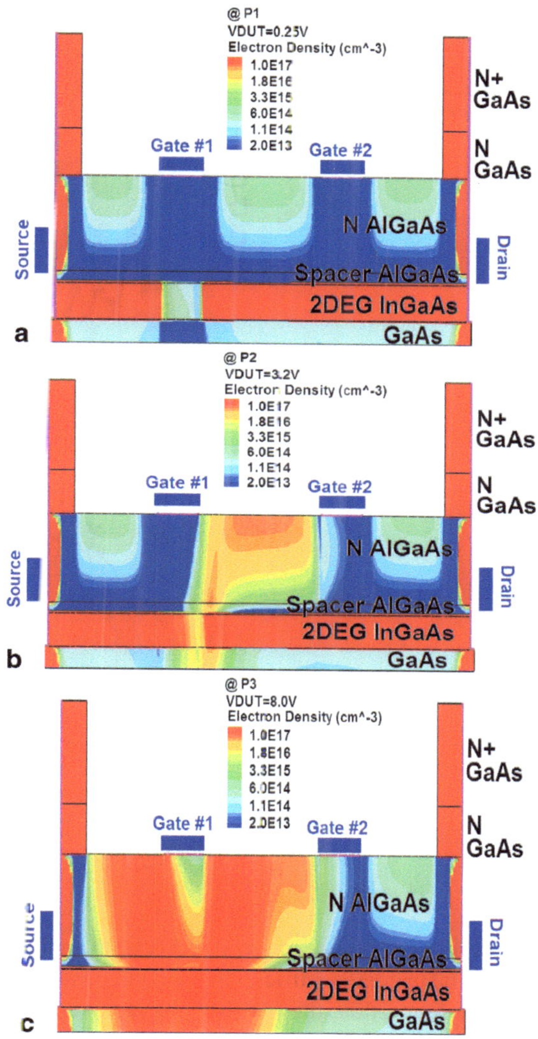

Fig. 4.16 Simulated electron densities in the dual-gate pHEMT at a stress voltage of 0.25 V (**a**), 3.2 V (**b**), and 8.0 V (**c**)

radation due to thermal effect [63]. Below this current level, SG1 does not saturate and operates in a region where the clamping voltage drops considerable when the current decreases. This is fairly consistent with the results given in Fig. 4.17b, which shows that SG1 carries a current of more than 0.5 A before 55 ns, thus resulting in a large clamping voltage. After 55 ns, the current in SG1 is smaller than 0.5 A and the clamping voltage suddenly drops. For DG1, the saturation region does not exist, and the clamping voltage vs. time waveform is fairly constant (see Fig. 4.17a). It is worth pointing out that the quasistatic nature of TLP I–V curves in Fig. 4.14a cannot be used to fully explain the voltage and current transients in Fig. 4.17a, b.

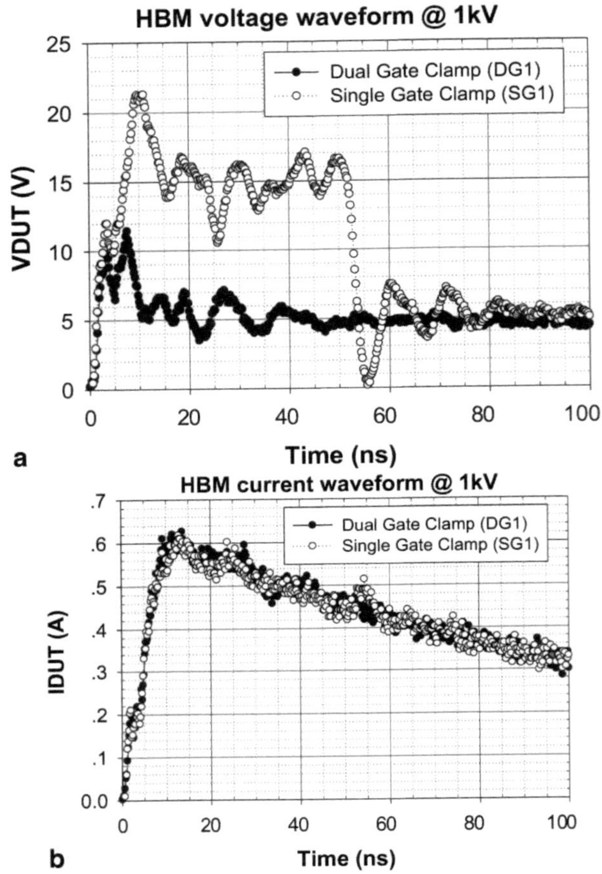

Fig. 4.17 Voltage (**a**) and current (**b**) waveforms of the two clamps subject to 1 kV HBM stress obtained from on-wafer HBM tester

For charged device model (CDM) ESD applications, the ESD protection devices have to be triggered sufficiently fast and possess a relatively low overshoot voltage. To investigate this, the pHEMT clamps were characterized using the Barth 4012 very-fast TLP (VFTLP) tester, which produces very fast and short pulses with a 100 ps rise time and 5 ns pulse width. Figure 4.18a shows the VFTLP I–V curves of SG1 and DG1, and Fig. 4.18b shows the voltage vs. time waveforms of the two devices measured using a VFTLP pulse with an amplitude of 37.5 V. Note that the failure currents obtained from the VFTLP measurement are quite similar to those from the TLP measurements. Our failure analysis revealed that the ESD-induced damages took place near the corner of the gate/N-AlGaAs interface, suggesting that the failure is field rather than thermal related. Moreover, it was found that the single-gate ESD clamp SG1 failed at a VFTLP pulse amplitude of 38 V, while the dual-gate clamp DG1 survived up to 94 V. We selected a stress voltage of 37.5 V to observe the voltage transients since neither DG1 nor SG1 was damaged and the ESD stress phenomenon was pronounced at such a stress condition. It can be seen

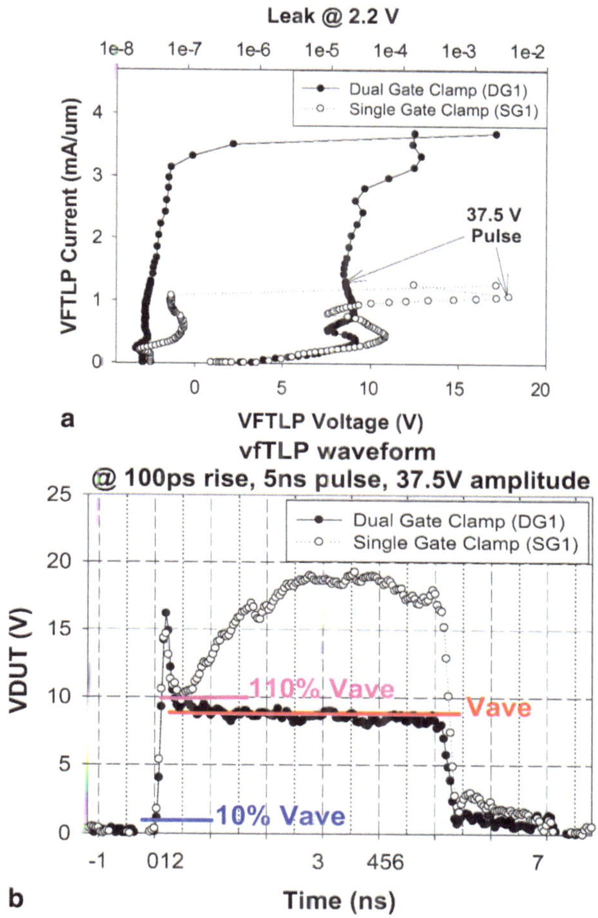

Fig. 4.18 **a** Current–voltage characteristics (the two curves on the *right-hand side*) and leakage currents (the two curves on the *left-hand side*) of SG1 and DG1 subject to VFTLP stresses. **b** Transient voltage waveforms of SG1 and DG1 subject to a VFTLP stress voltage of 37.5 V. Different Vave points for DG1 are also indicated

that both SG1 and DG1 have similar turn-on speeds and overshoot voltages. As such, for structures being protected and sensitive to overshoots, the ESD capabilities of SG1 and DG1 can be very similar. One the other hand, the dual-gate clamp has the advantage over the single-gate clamp of a much lower voltage after triggering. Specifically, after 0.5 ns, the transient voltage waveform of SG1 rises up again to a plateau region, while the voltage waveform of DG1 is kept at a fairly constant level (see Fig. 4.18b). This phenomenon can be explained as follows. Following the voltage overshoot, a large current in SG1 is forced to flow in the narrow 2DEG InGaAs channel (see Fig. 4.12a), resulting in an increase in the lattice temperature and consequently a decrease in the electron mobility. This in turn increases the resistance and hence the voltage drop in SG1 [63]. In the contrast, for DG1, the creation of the additional current path in the AlGaAs layer owing to the presence of second gate keeps the resistance and voltage drop in such a device relatively constant. This makes the dual-gate clamp a highly attractive candidate for CDM ESD protection at the input pin where the ESD-induced damage is caused mostly from high voltage CDM transient. The turn-on time is defined as the time it takes from 10 % Vave to 110 % Vave, where Vave for DG1 is a constant voltage calculated by averaging the voltages between 25 % and 75 % time periods. The values of Vave, 10 % Ave, and 110 % Ave of DG1 are indicated in Fig. 4.18b. For SG1, since the voltage rises again after 0.5 ns, Vave is chosen as the nadir voltage at 0.5 ns.

Finally, we measured the S-parameters of the two clamps with zero bias using the Agilent E8364C PNA Microwave Network Analyzer to determine the parasitic capacitances at different frequencies. Figure 4.19a shows the ESD clamp's

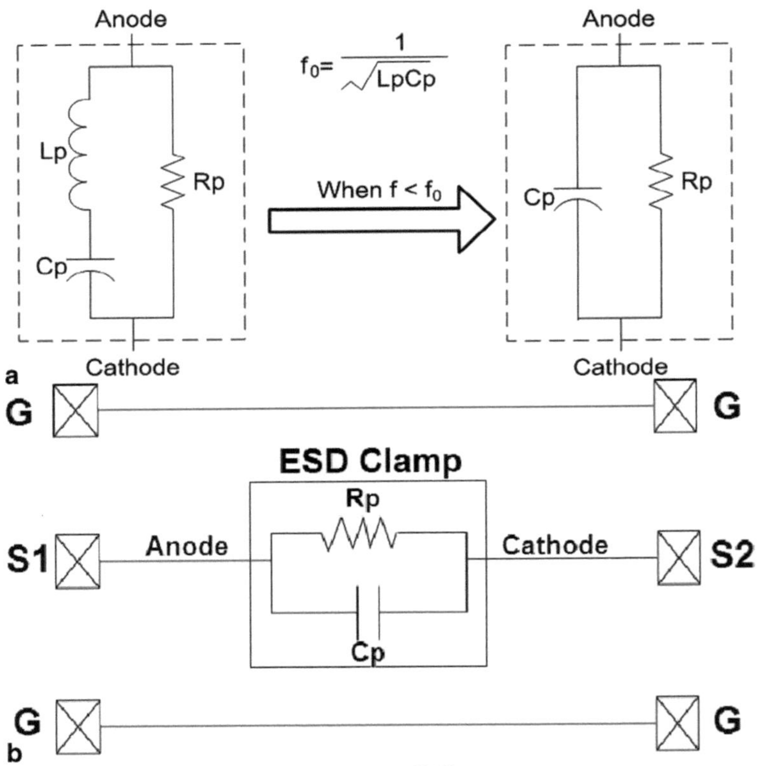

$$Y_{21} = \frac{-2S_{21}}{(1 + S_{11})(1 + S_{22}) - S_{21}S_{12}}$$

$$\mathrm{Cp} = \frac{\mathrm{Im}[-Y_{21}]}{2\pi f}$$

c

Fig. 4.19 **a** RLC model and simplified RC model. **b** Ground–signal–ground layout. **c** Capacitance equation for the pHEMT ESD clamp

equivalent RLC circuit consisting of a capacitor Cp, resistor Rp, and inductor Lp and its simplified Resistor-Capacitor (RC) model. In most cases, an ESD clamp is operating under frequencies lower than the resonant frequency f0, and the RLC model can be reduced to the RC model under such an operation. Figure 4.19b presents the ESD clamp's ground–signal–ground layout for S-parameter measurement, and Fig. 4.19c is the equation relating the capacitance to the S parameters based on the simplified RC model [66]. Figure 4.20a, b plot the capacitances normalized to the layout area (cap/area) and to the TLP failure current (cap/It2), respectively. The cap/area results of SG1 and DG1 are very similar, whereas DG1 shows notable

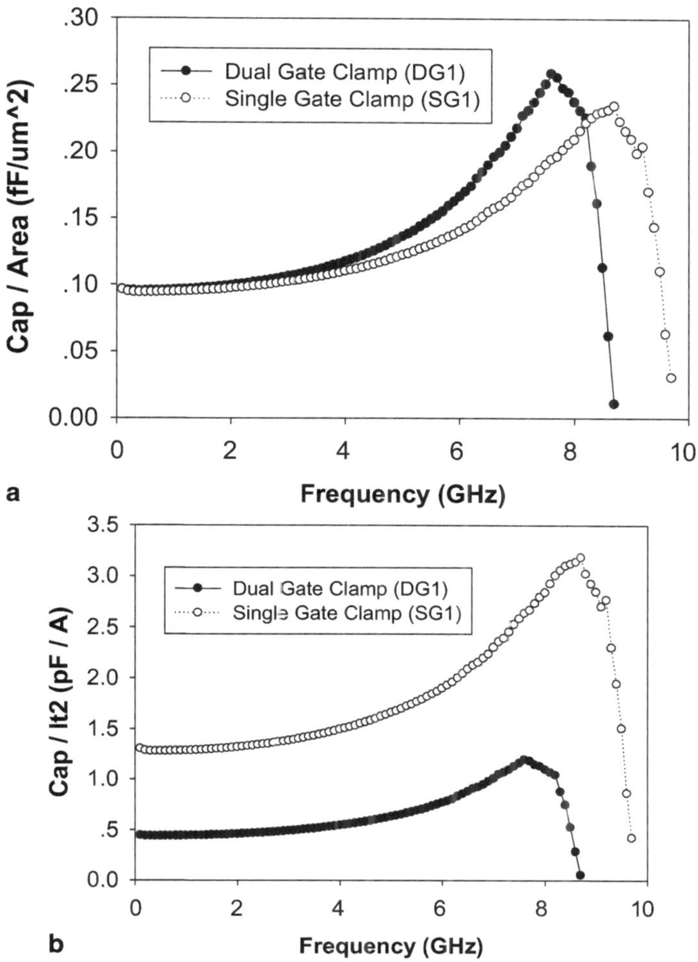

Fig. 4.20 Measured parasitic capacitances vs. frequency of two clamps normalized to area (**a**) and TLP failure current (**b**)

Table 4.4 Summary of ESD figure of merits (FOMs) for single-gate clamp (SG1) and dual-gate clamp (DG1)

ESD FOMs	SG1	DG1
pHEMT area (μm²)	116 × 106	135 × 106
Triggering diode area (μm²)	62 × 53	62 × 53
Current limiter resistor area (μm²)	40 × 20	40 × 20
Failure current I_{t2} (A)	1.15	3.80
Passing HBM (V)	2 K	> 8 K
Cap @ 5.2 GHz (pF)	1.95	2.47
Cap/Area @ 5.2 GHz (fF/μm²)	0.125	0.140
Cap/It2 @ 5.2 GHz (pF/A)	1.70	0.65
On-state resistance before failure (Ω)	14.0	3.07
Turn-on time (ns)	0.48	0.48
Leakage current (A)	2.7E−7	2.7E−7

advantageous over SG1 when cap/It2 is considered. When frequencies are higher than 8.5 GHz, both SG1 and DG1 exceed their resonate frequencies, and the devices turn from capacitive into inductive, causing the capacitance to drop sharply after this frequency.

The normalized capacitance of the single-gate pHEMT clamp reported in [62] was about 0.06 fF/μm² at 2 GHz, while both the single-gate/dual-gate pHEMT clamps in this paper have a normalized capacitance of about 0.1 fF/μm² at the same frequency. The difference is due mainly to the different processes used. Considering its superb failure current level, the proposed dual-gate clamp is still highly attractive for radio frequency (RF) ESD applications since a further trade-off can be made between the parasitic capacitance and failure current.

Table 4.4 summarizes the ESD performances of SG1 and DG1 pHEMT-based clamps.

The proposed dual-gate ESD clamp can readily be used as a power clamp for a pHEMT-based LNA with Vdd in the range of 2–2.5 V and Vss = 0, as the proposed ESD clamp has a trigger voltage of 3.2 V and a superb failure current. The device can also be used as an ESD protection device at the I/O pin, but the trigger voltage and parasitic capacitance will need to be optimized according to the I/O requirements.

4.3 Multigate D-Mode pHEMT-Based ESD Protection Clamp

The preceding section presented a novel E-mode dual-gate ESD clamp in the GaAs pHEMT process. This clamp has shown much better ESD protection ability than traditionally used stacked Schottky diode chain and single-gate pHEMT ESD clamp. However, such a clamp cannot be directly applied to the depletion mode (D-mode)-only GaAs pHEMT process because of the lack of E-mode pHEMT. An ESD protection solution built in the D-Mode HEMT technology is highly in demand because a lot of RF switches and RF power amplifiers are fabricated in the D-Mode HEMT technology [67]. In this section, we will investigate and analyze the effectiveness of multigate D-mode pHEMT-based ESD protection structures. As will be shown later, the dual-gate, D-Mode pHEMT ESD structure has a very attractive ESD discharging capability (failure current I_{t2} at 1.2 A) and the flexibility to adjust the trigger voltage for fulfilling different ESD protection design requirements.

4.3.1 Device Structure

Figure 4.21a, b shows the cross-sectional view of the proposed D-mode dual-gate GaAs pHEMT and schematic of ESD clamp built based on such a device. The ESD clamp consists of a trigger diode and a resistor (i.e., current limiter) to control the triggering of the pHEMT. Gate #1 in the dual-gate pHEMT is connected to the "trigger" terminal and Gate #2 is connected to the anode. Also note that several diodes (see Pinch-OFF Diode in the figure) are connected in series with pHEMT to turn it off under the normal operation since the D-Mode HEMT is a normal-on device. The pinch-off diode chain also helps in limiting the leakage current through the protection clamp. Under the normal operation condition, the trigger diode chain is not turned on, potential at the trigger terminal is low, pHEMT is off, and current flow between the anode and cathode is negligibly small.

Under a sufficiently large ESD stress, the voltage at the anode is larger than the blocking voltage of the diode string and subsequently turns on the diode chain and forces a current through the resistor. This raises the potential at the trigger terminal, reduces the depletion region underneath Gate #1 and Gate #2, and exposes the 2DEG channel and the second N-AlGaAs discharge path. More detailed explanations for the mechanisms of two current conducting paths will be given below.

Figure 4.21c presents the layout the dual-gate pHEMT-based clamp. It was fabricated in a 0.5-μm D-mode GaAs pHEMT technology at WIN Semiconductor. A triggering unit consisting of a reverse Schottky diode and a current limiter was incorporated in the circuit. Trigger diode area was $70 \times 40\ \mu m^2$, the current limiter resistor area was $64 \times 5\ \mu m^2$, pitch-off diode area was $180 \times 190\ \mu m^2$, and the dual-gate pHEMT had 16 fingers and areas of $180 \times 60\ \mu m^2$. The total effective ESD cell area is $300 \times 300\ \mu m^2$ including the pad areas. In some application, on could further squeeze the layout size by: (1) reducing the trigger diode area since it is just a trigger unit and does not need a large area to discharge ESD current and (2) removing the testing pad structures.

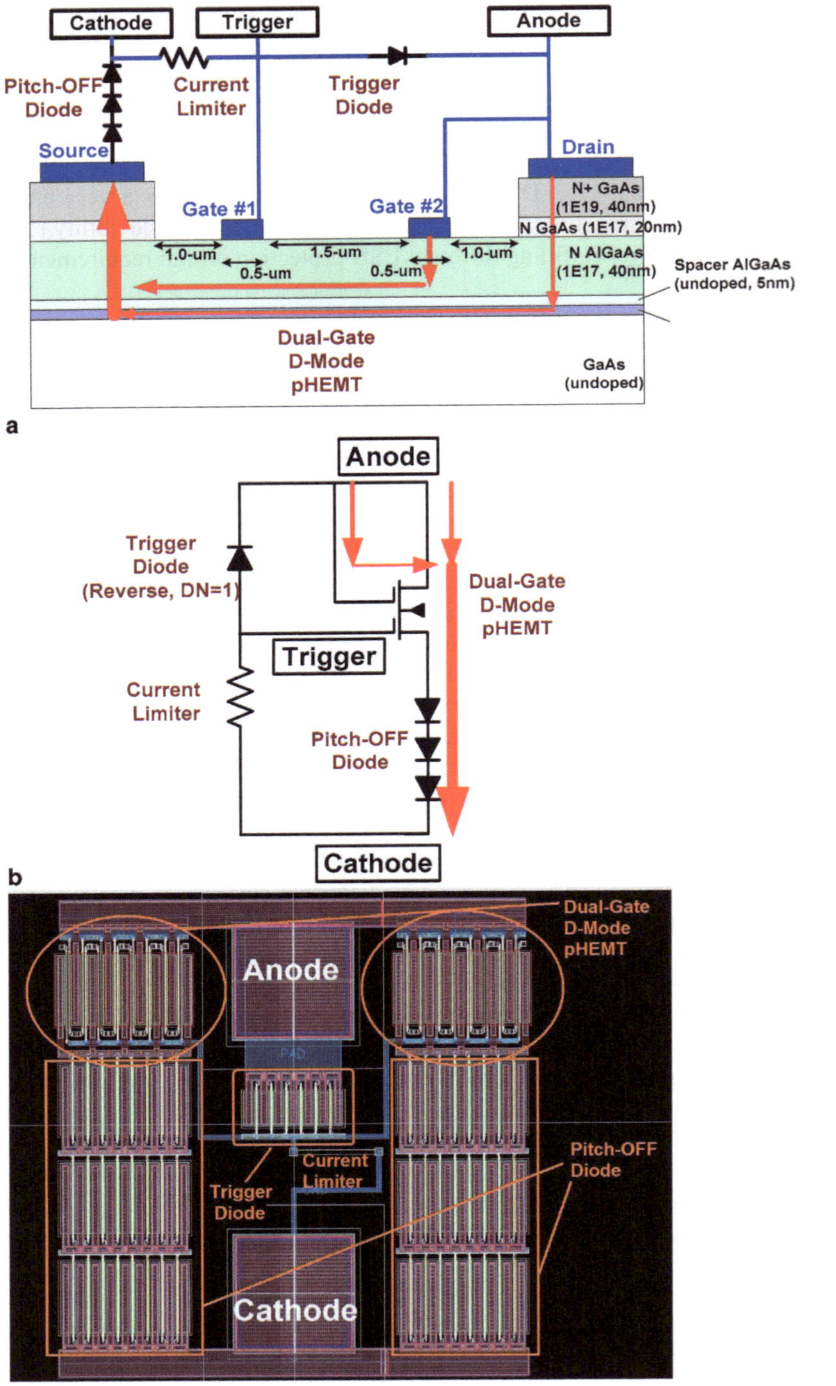

Fig. 4.21 **a** Cross-sectional view of proposed dual-gate ESD protection clamp built in D-Mode pHEMT technology with the two *red lines* denoting the two current paths. **b** Schematic view. **c** Layout of the ESD clamp

4.3.2 *Measurements and Discussion*

We performed HBM-like ESD measurements using the Barth 4002 TLP tester, which generates pulses having a 10 ns rise time and 100 ns width. Poststress leakage currents were measured at a voltage of 3.6 V. This voltage is the pHEMT normal operating voltage of 3.3 V, plus a 10% dynamic voltage margin.

Figure 4.22a presents the TLP current (y-axis) vs. voltage (bottom x-axis) and leakage current (top x-axis) curves of the dual-gate pHEMT ESD clamp. The failure current of the clamp is 1.16 A, which can be directly correlated to HBM passing voltages by conventional estimate of multiplying the failure current to 1500 Ω hu-

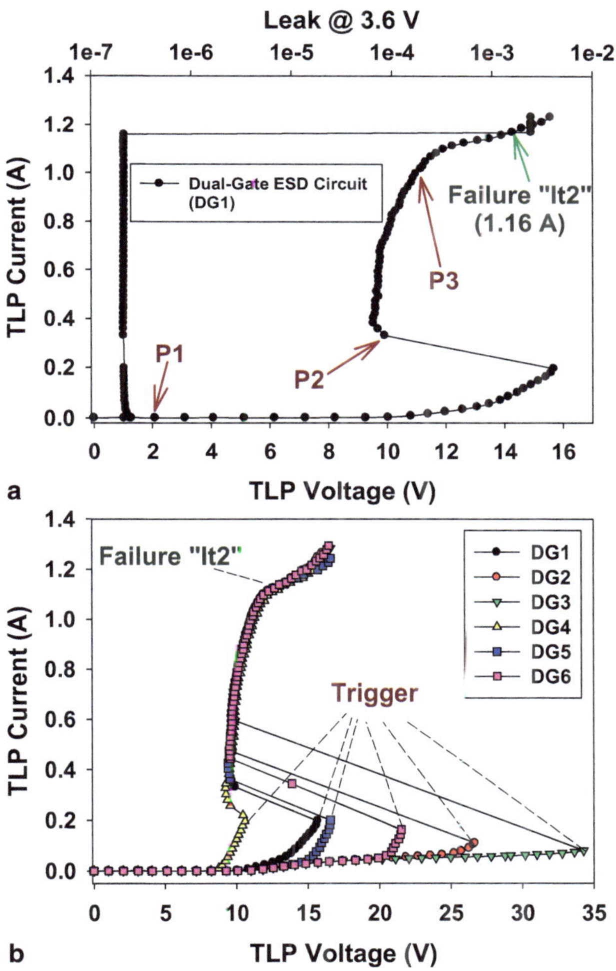

Fig. 4.22 **a** Current–voltage characteristics (on the *right-hand side*) and leakage currents (on the *left-hand side*) of dual-gate ESD protection circuit (DG1) having a reverse trigger diode from the TLP tester. **b** TLP results of dual-gate ESD circuits having different trigger diode number and polarity. (see DG1, DG2, DG3, DG4, DG5, and DG6 in Table 4.6)

man body resistance [68]. Thus, it is equivalent to an HBM passing level of 1.75 kV. We suspect the ESD damage occurred in the pinch-off diodes since the dual gate pHEMT (DG1) exhibits a high on-state resistance prior to its failure, which is similar to the failure of a stacked Schottky diode chain. Table 4.5 compares the TLP measurement results of the proposed dual-gate D-mode HEMT ESD circuit with previously reported single-gate D-mode clamp [62]. Clearly, the proposed dual-gate ESD circuit possesses a much higher failure current than its single-gate counterpart. Also, the proposed dual-gate ESD circuit has a better ESD robustness when normalized to the pHEMT area (0.162 vs. 0.12 A/μm²). Note that the trigger voltage depends on the number of trigger diodes, and this voltage can be adjusted by altering the diode number and diode polarity. The trigger voltage of about 15.68 V observed in Fig. 4.22a resulted from the use of a reverse diode plus three pinch-off diodes. Figure 4.22b shows the TLP I–V characteristics of dual-gate pHEMT's having six different trigger diode units. While the trigger voltage depends on the diode number, the failure current is insensitive to such a factor. Table 4.6 summarizes the measured trigger voltage as a function of diode number and polarity.

Referring to Fig. 4.23, the different current conduction mechanisms can be explained in more detail below using physical reasoning. When the stress voltage is at around 2 V (point P1 in Fig. 4.22a), the 2DEG channel underneath Gate #2 is unblocked but underneath Gate #1 is still obscured by the depletion region because of the blockade from the trigger diode and pinch-off diode chain (see schematic in Fig. 4.23a). When the ESD stress increases (point P2 in Fig. 4.22a), the Schottky diode begins to turn on but the current path from the Schottky contact to the source via the AlGaAs layer is still blocked by the depletion region associated with Gate #1 (see schematic in Fig. 4.23b). The 2DEG channel underneath Gate #1 is unveiled at this time because the trigger diode is conducting. So, there is only one current con-

Table 4.5 Comparison ESD results of proposed dual-gate ESD circuit (DG1) and single-gate ESD circuit [62]

ESD FOMs	Dual-gate "DG1"	Single gate in [62]
Failure current I_{t2} (A)	1.16	0.6
Equivalent HBM ESD level (V)	1.75 kV	0.87 kV
HEMT area (μm²)	180 × 60	90 × 80
ESD/HEMT area (V/μm²)	0.162	0.120

Table 4.6 Trigger voltage dependency on trigger diode number/polarity for D-Mode pHEMT ESD clamp

Dual-gate circuits	Diode polarity	Diode number	Trigger (V)	Failure "I_{t2}" (A)
DG1	Reverse	1	15.68	1.16
DG2	Reverse	2	26.67	1.17
DG3	Reverse	3	34.31	1.16
DG4	Forward	10	10.59	1.18
DG5	Forward	18	16.58	1.17
DG6	Forward	25	21.54	1.16

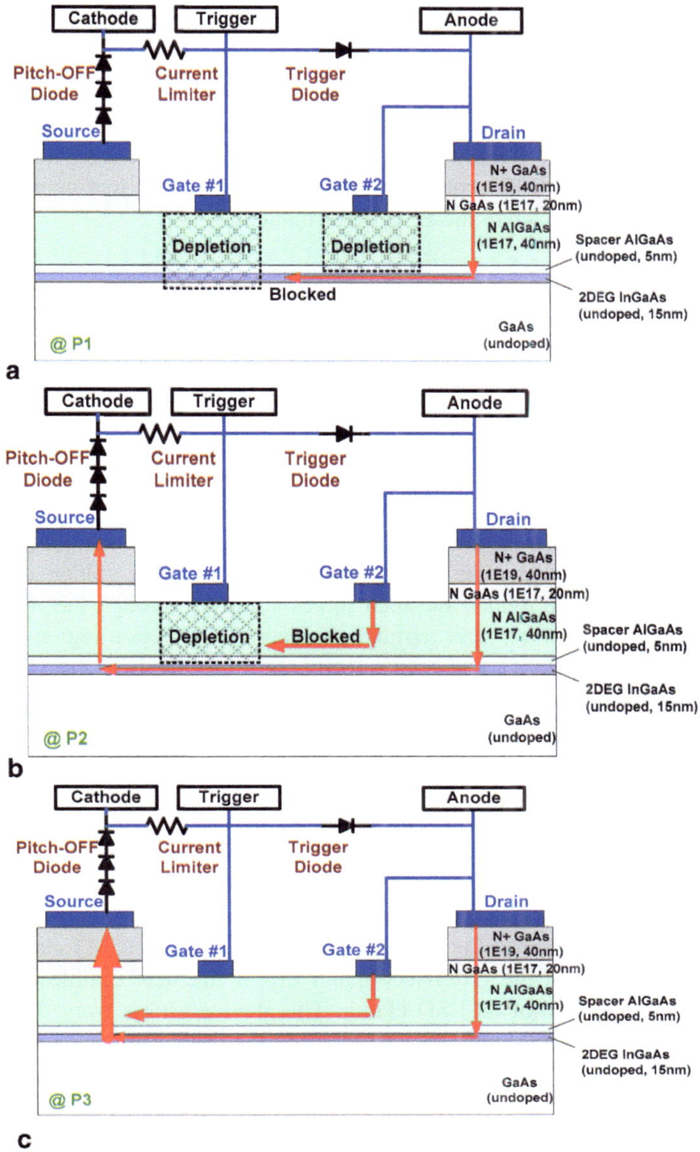

Fig. 4.23 Depletion regions and current paths in the dual-gate pHEMT at points (see Fig. 4.22) P1 (**a**), P2 (**b**), and P3 (**c**)

ducting path between the anode and cathode via the 2DEG channel (see Fig. 4.23b). As the stress voltage further increases (point P3 in Fig. 4.22a), the depletion region associated with Gate #1 also vanishes, and the second current path associated with the Schottky diode is formed (see Fig. 4.23c). As a result, two current conducting paths exist in the dual-gate pHEMT, which gives rise to a high robustness for the dual-gate D-mode pHEMT ESD protection structure.

4.4 Drain-Less pHEMT-Based ESD Protection Clamp

In the previous sections, dual-gate GaAs pHEMT-based ESD clamps demonstrating an excellent ESD current handling ability were developed [69, 70]. To further enhance the performance, in this section we will propose and realize a new drainless, multigate pHEMT for ESD protection applications. TLP measurements will again be carried out to characterize the ESD performances.

4.4.1 Device Structure

Figure 4.24a shows the cross section view of a conventional E-mode, single-gate GaAs pHEMT and schematic of ESD clamp built based on such a device [62, 71]. The ESD clamp consists of a diode chain and a resistor (i.e., current limiter) to control the triggering of the pHEMT. Under the normal operation condition, the diode chain is not turned on, the potential at the trigger terminal is low, the pHEMT is off, and the current flow between the anode and cathode is negligibly small. Under the ESD condition, a stress voltage at the anode larger than the blocking voltage of the diode string first turns on the diodes and then forces a current through the resistor. This raises the potential at the trigger terminal, reduces the depletion region underneath the gate, exposes the 2DEG, and consequently allows for the ESD current to be discharged between the anode and cathode via the 2DEG channel (indicated by the red line in Fig. 4.24a). Figure 4.24b shows the cross-sectional view of a dual-gate GaAs pHEMT and schematic of its ESD clamp [70]. Gate #1 in the dual-gate pHEMT is the same gate in the single-gate pHEMT, but the dual-gate pHEMT has a second gate, Gate #2, which is connected to the anode. Subjecting to a sufficiently large ESD stress voltage, the Schottky junction underneath Gate #2 will conduct. This forms a second current discharging path via the AlGaAs layer (see Fig. 4.24b), in addition to the 2DEG current path created by Gate #1. The added current path reduces the on-state resistance and increases the robustness of the dual-gate pHEMT clamp.

Figure 4.24c shows the cross-sectional view of the new drainless, multigate pHEMT and schematic of its ESD clamp. This device differs from the dual-gate GaAs pHEMT in the following two aspects: (1) the drain region is removed considering the fact that it contributes minimally to the ESD current discharge while it does consume a considerable layout area [70], and (2) the structure is symmetrically mirrored, with two D-mode gates connected to the anode terminal, two E-mode gates connected to the trigger terminal, and two source terminals connected to the cathode terminal for better current distribution uniformity. Under a sufficiently large ESD stress voltage, the Schottky junction underneath the D-mode Gate #2 and Gate #4 will conduct. This forms a symmetrical a current discharge path from Gate #2/Gate #4 to the source via the AlGaAs layer (see Fig. 4.24c, in addition to the 2DEG current path created by Gate #1/Gate #3. The added current path reduces the on-state resistance and increases the robustness of the drainless, multigate pHEMT ESD clamp. The reason underlying the higher ESD robustness in the drainless

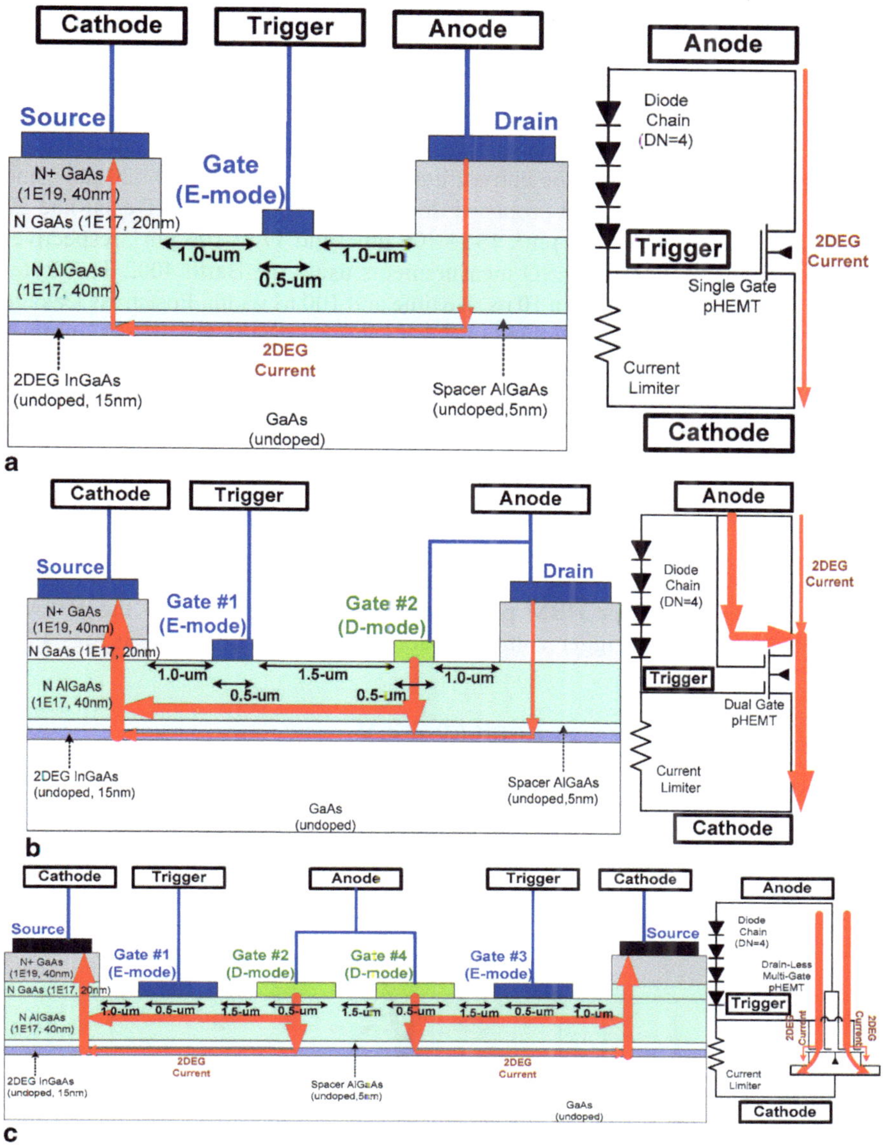

Fig. 4.24 Cross-sectional view and respective ESD clamp circuit for single-gate pHEMT (**a**), **b**dual-gate pHEMT (**b**), and drainless, multigate pHEMT (**c**)

multigate pHEMT clamp is the fact that the conduction-inefficient "drain" in the conventional pHEMT is now replaced with a symmetrical, highly conductive channel in the AlGaAs layer created by the two D-mode gates, while the traditional current path via the 2DEG is still maintained (see Fig. 4.24c).

4.4.2 Measurements and Discussion

Single-gate, dual-gate, and new drainless multigate E-mode pHEMT clamps were fabricated in a 0.5-μm pHEMT process at WIN Semiconductor. A triggering unit consisting of a four-diode chain and a current limiter was incorporated in the clamps. Diode area was 62 × 53 μm², the current limiter resistor area was 40 × 20 μm², and the single-gate, dual-gate, and proposed drainless multigate pHEMT each had 20 fingers and areas of 116 × 106 μm², 135 × 106 μm², and 175 × 106 μm², respectively. We performed HBM-like ESD measurements using the Barth 4002 TLP tester which generates pulses having a 10 ns rise time and 100 ns width. Poststress leakage currents were measured at a voltage of 2.2 V. This voltage is the pHEMT normal operating voltage of 2.0 V, plus a 10 % dynamic voltage margin.

Figure 4.25 presents the TLP current (*y*-axis) vs. voltage (bottom *x*-axis) and leakage current (top *x*-axis) curves of the single-gate, dual-gate, and drainless multigate ESD clamps. The measurement results were also summarized in Table 4.7. Clearly, the drainless multigate pHEMT clamp has the highest failure current I_{t2} of 5.2 A (HBM passing voltage of 7.8 kV), followed by the dual-gate pHEMT clamp with I_{t2} of 2.0 A (HBM passing voltage of 3.0 kV), and the traditional single-gate pHEMT clamp has the lowest failure current of only 1.0 A. Moreover, the drainless clamp also has the highest HBM passing level when normalized by the pHEMT area (see Table 4.7). A trigger voltage of about 3.5 V was observed in Fig. 4.25,

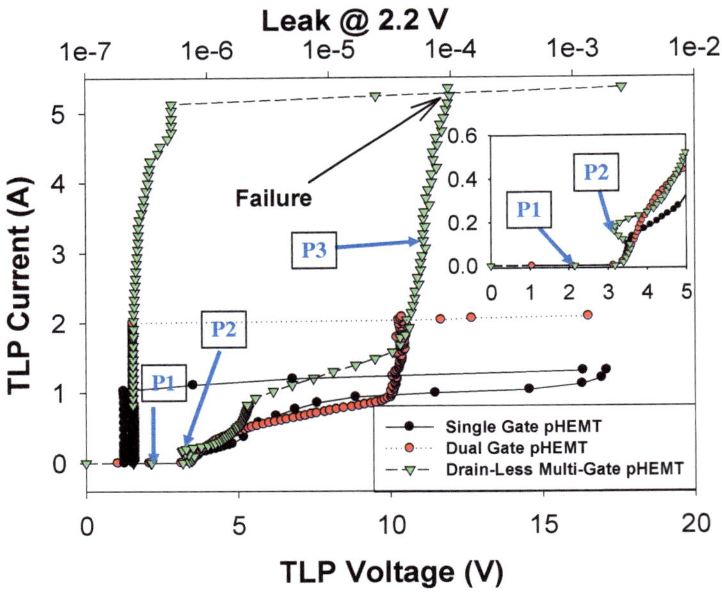

Fig. 4.25 Current–voltage characteristics (the three curves on the *right-hand side*) and leakage currents (the three curves on the *left-hand side*) of single-gate pHEMT clamp, dual-gate clamp, and drainless multigate pHEMT clamp having a four-diode chain obtained from the TLP tester

Table 4.7 Failure currents and areas of single-gate pHEMT, dual-gate pHEMT, and drainless multigate pHEMT ESD clamps

Device	Failure current "It2" (A)	pHMET device area (μm²)	ESD/Area (V/μm2)
Single gate	1.1	116 × 106	0.1342
Dual gate	2.0	136 × 106	0.2081
Four gate	5.2	175 × 106	0.4205

resulting from the use of a four-diode chain, and this voltage can be adjusted by altering the diode number and diode polarity [70]. The proposed drainless multigate pHEMT clamp can readily be used as a power clamp for a pHEMT-based LNA with a supply voltage range of 2–3 V.

The current conduction mechanisms can be explained in more detail in Fig. 4.26. When the stress voltage is at around 2 V (point P1 in Fig. 4.25), the 2DEG channels underneath Gate #2/Gate #4 are unblocked but underneath Gate #1/Gate #3 are

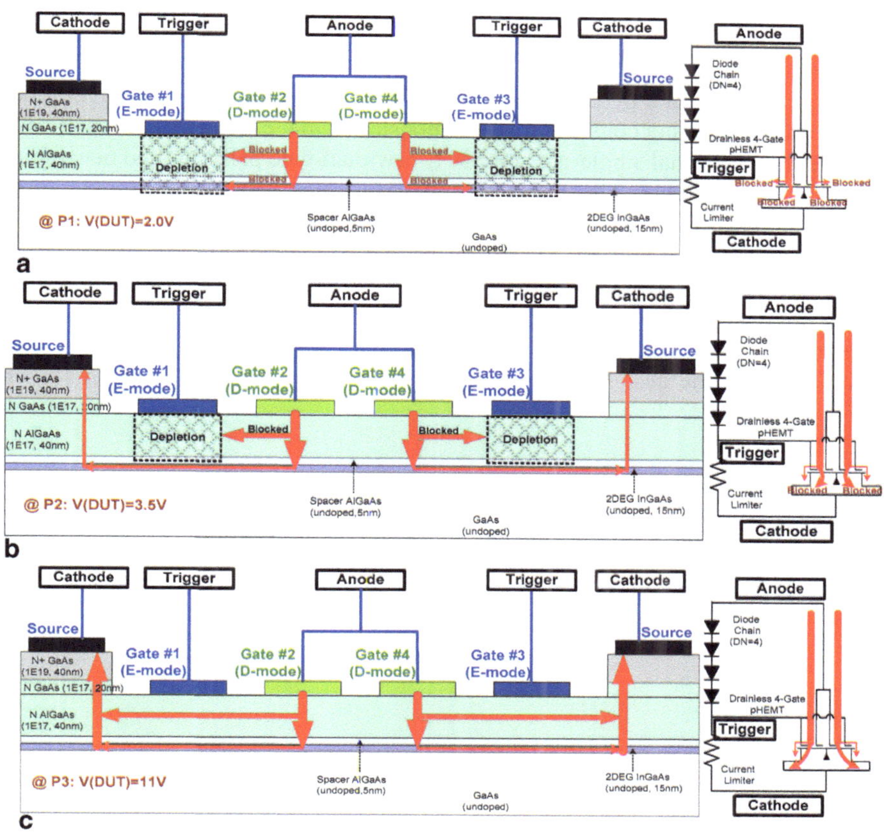

Fig. 4.26 Depletion regions and current paths in the drainless multigate pHEMT ESD clamp at points (see Fig. 4.25) P1 (**a**), P2 (**b**), and P3 (**c**)

still obscured by the depletion region because of the blockade of diode chain (see schematic in Fig. 4.24c). When the ESD stress increases (P2 in Fig. 4.25), the diode chain begins to turn on but the current paths via the AlGaAs layer are still blocked by the depletion region (see schematic in Fig. 4.26b). The 2DEG channels underneath Gate #1/Gate #3 are unveiled at this time because the diode chain is conducting. So there is only two current conducting paths between the anode and cathode via the 2DEG channel (see Fig. 4.26b). As the stress voltage further increases (point P3 in Fig. 4.25), the depletion regions in the AlGaAs layer associated with Gate #1/ Gate #3 also vanish, and the second current paths associated with the diode chain are formed (see Fig. 4.26c). As a result, there are four symmetrical current conducting paths in the proposed drainless multigate pHEMT, thus giving rise to the much larger failure current.

4.5 Summary

In this chapter, several GaAs pHEMT-based ESD protection structures were developed, analyzed, and discussed. While these structures consist generally of the same components of a pHEMT, a resistor, and a number of diodes, different types of pHEMT were being considered and demonstrated in the chapter. There pHEMTs include the traditional single-gate pHEMT, new dual-gate pHEMT, and new drainless multigate pHEMT. The single-gate pHEMT-based ESD clamp is superior to the widely used Schottky diode chain, but its robustness is still quite poor. Due to the availability of multiple current paths, the dual-gate and drainless pHEMT-based clamps showed significantly improved ESD performances over the conventional single-gate counterpart, including a higher current discharge capability, lower on-state resistance, and smaller voltage transient. These results provided useful information on the design of effective ESD protection solutions for GaAs-based integrated circuits.

Chapter 5
Conclusion

Advancement in radio frequency (RF) integrated circuits (ICs) is one of the driving forces behind the wide spread use of modern wireless communication and other high-speed applications. The reliable daily usage of these RF electronics cannot be assured, however, unless the issue of electrostatic discharge (ESD) protection is properly addressed and implemented. As the semiconductor technology continues to scale down, ESD reliability concern is becoming more and more pervasive due to the shrink of device size and reduction of breakdown voltage. In order to mitigate the ESD-induced IC failures, on-chip ESD protection devices are commonly employed to discharge ESD-induced current and to limit the voltages at IC pins to a satisfactorily low level during ESD events.

Effective ESD devices aimed at protecting low-speed analog and digital circuits fabricated in CMOS processes have been extensively reported in the literature. On the contrary, a text focuses exclusively on ESD protection devices for RF ICs is not yet available, and this book was written to fill such a critical void.

ESD protection design of RF ICs is way more challenging than that of traditional low speed counterpart because of the facts that: (1) high-performance RF ICs are typically fabricated in a compound semiconductor process such as the GaAs pHEMT or SiGe HBT technology, but some typical ESD devices are not able to be realized in such a process; (2) compound semiconductor has a relatively low thermal conductivity which can worsen the ESD protection capability; (3) inherent parasitic capacitances associated with ESD protection devices can degrade significantly the RF functionality of the core circuit, and hence the transparency of RF ESD devices is a major consideration.

The book began with the introduction of ESD phenomena fundamentals, ESD models, ESD test methods, and different types of ESD device's characteristics. For RF ICs, the following three devices can typically be used for building ESD protection solutions: (1) diode and diode chain, (2) externally triggered field-effect transistor, and (3) silicon-controlled rectifier (SCR). However, each of them has its own shortcomings which prevent any of them from being dominating the RF IC ESD applications. Diode chain has a large layout area, large on-state resistance and poor leakage under the "OFF" state. For the case of the externally triggered field-effect

Q. Cui et al., *On-Chip Electro-Static Discharge (ESD) Protection For Radio-Frequency Integrated Circuits*, DOI 10.1007/978-3-319-10819-3_5

transistor, for example, an externally triggered GaAs pHEMT has only limited ESD protection capability due to the fact that there is only one current conducting path (via the 2DEG channel). SCR could be fabricated in a SiGe BiCMOS process and standard CMOS process, but the SCR is relatively slow in turning on and is prone to ESD latchup. Motivated by the above-mentioned design challenges associated with the RF ESD, this book provided a systematical and comprehensive coverage on the ESD protection design and analysis for dominating RF technologies including the GaAs pHEMT, SiGe BiCMOS, and Si CMOS processes.

With the above-mentioned device physics described and understood, Chap. 2 covered the development of an effective ESD protection solution for Si CMOS RF ICs. Among the various ESD devices, SCR is considered the best candidate due to its excellent ESD robustness. However, the parasitic capacitance of SCR needs to be reduced to increase the SCR's transparency and thus to mitigate the SCR's parasitic effect on RF performance. To this end, a novel SCR-based ESD structure was introduced and characterized for ESD protection of RF CMOS based integrated circuits. A theoretical analysis was also given to provide the physics underlying the low capacitance observed in the new ESD protection device.

The focus of Chap. 3 was ESD protection in SiGe technology. SiGe BiCMOS is another important compound semiconductor process for RF ICs, and the SiGe-based SCR is a good candidate for ESD protection application in this area. However, the SiGe SCR has a slow turn-on speed issue when it is used for a relatively fast ESD event like the charged device model (CDM ESD). In order to optimize the turn-on performance, a new SiGe SCR with a P PLUG layer was developed. Both the very fast TLP (vfTLP) measurements and vfTLP-like TCAD simulations were carried out for characterization and analysis.

Finally, in Chap. 4, conventional and novel GaAs pHEMT-based ESD protection structures were introduced. For example, multigate pHEMT-based ESD protection devices in both the enhancement- and depletion-modes were developed and characterized. Due to the multiple current paths available in these devices, the new ESD protection clamp could offer significantly improved ESD performances over the conventional single-gate pHEMT ESD clamp, including a higher current discharge capability, lower on-state resistance, and smaller transient voltage. Moreover, this chapter presented an enhanced ESD protection clamp based on a novel drain-less, multigate pHEMT. The proposed ESD protection clamp can possess an even higher current-handling capability than the multigate pHEMT clamp.

The materials contained in this book should offer useful and timely information on the subject of RF ESD for the device and circuit engineers alike, as well as students and researchers who are involved in the general fields of microelectronics and its reliability.

References

1. Amerasekera A, Ramaswamy S, Mi-Chang C, Duvvury C. Modeling MOS snapback and parasitic bipolar action for circuit-level ESD and high current simulations. In: Reliability Physics Symposium, 1996. 34th Annual Proceedings. IEEE International. 1996, pp. 318–26.
2. Joshi S, Ida R, Rosenbaum E. Design and optimization of vertical SiGe thyristors for on-chip ESD protection. IEEE Trans Device Mater Reliab. 2004;4:586–93.
3. Ker MD, Hsiao YW, Wu WL. ESD-protection design with extra low-leakage-current diode string for RF circuits in SiGe BiCMOS process. IEEE Trans Device Mater Reliab. 2006;6:517–27.
4. Maloney TJ, Dabral S. Novel clamp circuits for IC power supply protection. IEEE Trans Compon Packag Manuf Technol Part C 1996;19:150–61.
5. Ming-Dou K, Kun-Hsien L. Double snapback characteristics in high-voltage nMOSFETs and the impact to on-chip ESD protection design. IEEE Electron Device Letters. 2004;25:640–2.
6. Rosenbaum E, Hyvonen S. On-chip ESD protection for RF I/Os: devices, circuits and models. In: IEEE International Symposium on Circuits and Systems, 2005. ISCAS 2005, vol. 2. 2005, pp. 1202–5.
7. Vashchenko VA, Hopper P. Bipolar SCR ESD devices. Microelectron Reliab. 2005;45: 457–71.
8. Voldman SH. A review of electrostatic discharge (ESD) in advanced semiconductor technology. Microelectron Reliab. 2004;44:33–46.
9. Wang AZH, Haigang F, Rouying Z, Haolu X, Guang C, Qiong W, Guan X, Zhihua W, Chun Z. A review on RF ESD protection design. IEEE Trans Electron Devices. 2005;52:1304–11.
10. Vinson JE, Liou JJ. Electrostatic discharge in semiconductor devices: an overview. Proc IEEE. 1998;86:399–420.
11. Brandt M, Halperin S. What does ESD really cost? Circuits assembly magazine. 2003. http://www.circuitsassembly.com/ca/index.php/magazine/24540-component-processing-1501; http://circuitsassembly.com/cms/images/stories/pdf/0306/0306esd.pdf.
12. Bo-Jr H, Chi-Hsueh W, Chung-Chun C, Ming-Fong L, Pin-Cheng H, Kun-You L, Huei W. Design and analysis for a 60-GHz low-noise amplifier with RF ESD protection. IEEE Trans Microw Theory Tech. 2009;57:298–305.
13. Ming-Dou K, Chun-Yu L. Test structure on SCR Device in waffle layout for RF ESD protection. In: IEEE International Conference on Microelectronic Test Structures, 2007. ICMTS '07. 2007, pp. 196–9.
14. D. o. Defense, MIL-STD-883E, test method standard: microcircuits, 1996.
15. Discharge E. Sensitivity testing Machine Model (MM), 1997. EIA/JEDEC Standard EIA, JESD22-A115–A.
16. ANSI/ESD, STM5.2-2012 Electrostatic discharge sensitivity testing—Machine Model (MM)—Component Level 2012.

Q. Cui et al., *On-Chip Electro-Static Discharge (ESD) Protection For Radio-Frequency Integrated Circuits,* DOI 10.1007/978-3-319-10819-3

17. IEC, IEC 60749-27 Semiconductor devices—Mechanical and climatic test methods—Part 27: Electrostatic Discharge (ESD) sensitivity testing—Machine Model (MM), 2003.
18. AEC-Q100-003-Rev-E, Machine Model (MM) Electrostatic discharge tes, 2012.
19. Peirce RJ. ESD-coping with ESD—the most common causes of ESD damage. Eval Eng. 2002;41:56–9.
20. Amerasekera EA, Duvvury C, Anderson W, Gieser H, Ramaswamy S. ESD in silicon integrated circuits. Chichester: Wiley; 2002.
21. Maloney TJ. Evaluating TLP transients and HBM waveforms. In: 31st EOS/ESD Symposium, 2009, 2009, pp. 1–9.
22. Ker MD. Whole-chip ESD protection design with efficient VDD-to-VSS ESD clamp circuits for submicron CMOS VLSI. IEEE Trans Electron Devices. 1999;46:173–83.
23. Amerasekera A, Ramaswamy S, Chang MC, Duvvury C. Modeling MOS snapback and parasitic bipolar action for circuit-level ESD and high current simulations. In: Reliability Physics Symposium, 1996. 34th Annual Proceedings. IEEE International. 1996, pp. 318–26.
24. Ker MD, Lo WY. Design on the low-leakage diode string for using in the power-rail ESD clamp circuits in a 0.35-/spl mu/m silicide CMOS process. IEEE J Solid-State Circuits. 2000;35:601–11.
25. Muthukrishnan S, Iversen C, Peachey N. A novel on-chip protection circuit for RFICs implemented in D-mode pHEMT technology. In: 29th Electrical Overstress/Electrostatic Discharge Symposium. p. A, 2007, pp. 3–7.
26. Natarajan MI, Thijs S, Jansen P, Tremouilles D, Jeamsaksiri W, Decoutere S, Linten D, Nakaie T, Sawada M, Hasebe T. RFCMOS ESD protection and reliability. In: Proceedings of the 12th International Symposium on the Physical And Failure Analysis of Integrated Circuits, 2005. IPFA 2005. 2005, pp. 59–66.
27. Thijs S, Natarajan MI, Daenen T, Degraeve R, Linten D, Wambacq P, Vassilev V, Groeseneken G, Scholten A. 1B. 2 ESD Protection for 5.5 GHz LNA in 90 nm RF CMOS-Implementation Concepts, Constraints and Solutions. In: Electrical Overstress Electrostatic Discharge Symposium Proceedings. 2004, p. 40.
28. Leroux P, Steyaert M. High-performance 5.2 GHz LNA with on-chip inductor to provide ESD protection. Electron Lett. 2001;37:467–9.
29. Ming-Dou K, Yuan-Wen H, Woei-Lin W. ESD-protection design with extra low-leakage-current diode string for RF circuits in SiGe BiCMOS process. IEEE Trans Device Mater Reliab. 2006;6:517–27.
30. Pierco R, Li Z, Torfs G, Yin X, Bauwelinck J, Qiu XZ. Diode string with reduced clamping-voltage for ESD-protection of RF-circuits. Electron Lett. 2012;48:317–8.
31. Guang C, Haigang F, Haolu X, Rouying Z, Qiong W, Guan X, Wang A, Takasuka K, Tamura S, Zhihua W, Chun Z. Characterizing diodes for RF ESD protection. IEEE Electron Device Lett. 2004;25:323–5.
32. Haolu X, Haigang F, Rouying Z, Wang A, Rodriguez D, Rice D. A new low-parasitic polysilicon SCR ESD protection structure for RF ICs. IEEE Electron Device Lett. 2005;26:121–3.
33. Ming-Dou K, Chun-Yu L. Low-capacitance SCR with waffle layout structure for on-chip ESD protection in RF ICs. IEEE Trans Microw Theory Tech. 2008;56:1286–94.
34. Liou JJ. Advanced semiconductor device physics and modeling. Boston: Artech House; 1993.
35. Lee Jian-Hsing, Wu YH, Peng KR, Chang RY, Yu TL, Ong T. The embedded SCR NMOS and low capacitance ESD protection device for self-protection scheme and RF application. In: Custom Integrated Circuits Conference. Proceedings of the IEEE. 2002, p. 93.
36. Salcedo JA, Liou JJ, Bernier JC. Design and integration of novel SCR-based devices for ESD protection in CMOS/BiCMOS technologies. IEEE Trans Electron Devices. 2005;52:2682–9.
37. Chun-Yu L, Ming-Dou K. Low-capacitance SCR with waffle layout structure for on-chip ESD protection in RF ICs. In: Radio Frequency Integrated Circuits (RFIC) Symposium, 2007 IEEE. 2007, pp. 749–52.
38. Voldman S. ESD: physics and devices. Hoboken: Wiley; 2004.

39. Mergens MPJ, Russ CC, Verhaege KG, Armer J, Jozwiak PC, Mohn R, Keppens B, Trinh CS. Diode-triggered SCR (DTSCR) for RF-ESD protection of BiCMOS SiGe HBTs and CMOS ultra-thin gate oxides. In: Electron Devices Meeting, 2003. IEDM '03 Technical Digest. IEEE International. 2003, pp. 21.3.1–4.
40. Ming-Dou K, Kuo-Chun H. SCR device fabricated with dummy-gate structure to improve turn-on speed for effective ESD protection in CMOS technology. IEEE Trans Semicond Manuf. 2005;18:320–7.
41. Mergens MPJ, Russ CC, Verhaege KG, Armer J, Jozwiak PC, Mohn RP, Keppens B, Trinh CS. Speed optimized diode-triggered SCR (DTSCR) for RF ESD protection of ultra-sensitive IC nodes in advanced technologies. IEEE Trans Device Mater Reliab. 2005;5:532–42.
42. Chen JZ, Amerasekera A, Vrotsos T. Bipolar SCR ESD protection circuit for high speed submicron bipolar/BiCMOS circuits. In: Electron Devices Meeting, 1995. IEEE International. 1995, pp. 337–40.
43. Parthasarathy S, Salcedo JA, Hajjar JJ. Design of SCR devices for SiGe BiCMOS applications. In: Bipolar/BiCMOS Circuits and Technology Meeting (BCTM). 2011, pp. 235–8.
44. Salcedo JA, Liou JJ, Liu Z, Vinson JE. TCAD methodology for design of SCR devices for electrostatic discharge (ESD) applications. IEEE Trans Electron Devices. 2007;54:822–32.
45. Gauthier R, Abou-Khalil M, Chatty K, Mitra S, Junjun L. Investigation of voltage overshoots in diode triggered silicon controlled rectifiers (DTSCRs) under very fast transmission line pulsing (VFTLP). In: 31st EOS/ESD Symposium, 2009. 2009, pp. 1–10.
46. Chu C, Gallerano A, Watt J, Hoang T, Tran T, Chan D, Wong W, Barth J, Johnson M. Using VFTLP data to design for CDM robustness. In: 31st EOS/ESD Symposium, 2009. 2009, pp. 1–6.
47. Gieser H, Haunschild M. Very-fast transmission line pulsing of integrated structures and the charged device model. In: Electrical Overstress/Electrostatic Discharge Symposium, 1996. Proceedings. 1996, pp. 85–94.
48. Esmark K, Gossner H, Bychikhin S, Pogany D, Russ C, Langguth G, Gornik E. Transient behavior of SCRS during ESD pulses In: Reliability Physics Symposium, 2008. IRPS 2008. IEEE International. 2008, pp. 247–53.
49. Di Sarro J, Chatty K, Gauthier R, Rosenbaum E. Study of design factors affecting turn-on time of silicon controlled rectifiers (SCRS) in 90 and 65 nm bulk CMOS technologies. In: Reliability Physics Symposium Proceedings, 2006. 44th Annual. IEEE International. 2006, pp. 163–8.
50. Vashchenko VA, Hopper PJ. New dual-direction ESD device in Si-Ge BiCMOS process. In: 10th IEEE International Conference on Solid-State and Integrated Circuit Technology (ICSICT). 2010, pp. 935–7.
51. Linten D, Roussel P, Scholz M, Thijs S, Griffoni A, Sawada M, Hasebe T, Groeseneken G. Calibration of very fast TLP transients. In: 31st EOS/ESD Symposium, 2009. 2009, pp. 1–6.
52. Qiang C, Yan H, Liou JJ, Shurong D. Analyse of protection devices' speed performance against ESD under CDM using TCAD. In: IEEE Conference on Electron Devices and Solid-State Circuits, 2007. EDSSC 2007. 2007, pp. 477–80.
53. Wybo G, Verleye S, Van Camp B, Marichal O. Characterizing the transient device behavior of SCRs by means of VFTLP waveform analysis. In: 29th Electrical Overstress/Electrostatic Discharge Symposium, 2007. EOS/ESD, 2007, pp. 6A.5-1–10.
54. Liou JJ. Principles and analysis of AlGaAs/GaAs heterojunction bipolar transistors. Boston: Artech House; 1996.
55. Lin CY, Chu LW, Ker MD. ESD protection design for 60-GHz LNA with inductor-triggered SCR in 65-nm CMOS process. IEEE Trans Microw Theory Tech. 2012;60:714–23.
56. Raczkowski K, Thijs S, Tseng JC, Chang TH, Song MH, Linten D, Nauwelaers B, Wambacq P. 60 GHz low noise amplifiers with 1 kV CDM protection in 40 nm LP CMOS. In: 2012 IEEE 12th Topical Meeting on Silicon Monolithic Integrated Circuits in RF Systems (SiRF). 2012, pp. 9–12.

57. Thijs S, Raczkowski K, Linten D, Tseng JC, Chang TH, Song MH, Groeseneken G. CDM protection for millimeter-wave circuits. In: 33rd Electrical Overstress/Electrostatic Discharge Symposium (EOS/ESD). 2011, pp. 1–7.
58. Chang T, Chen J, Rigge LA, Lin J. ESD-protected wideband CMOS LNAs using modified resistive feedback techniques with chip-on-board packaging. IEEE Trans Microw Theory Tech. 2008;56:1817–26.
59. Linten D, Natarajan MI, Thijs S, Van Huylenbroeck S, Xiao S, Carchon G, Decoutere S, Sawada M, Hasebe T, Groeseneken G. Implementation of 6 kV ESD Protection for a 17 GHz LNA in 130 nm SiGeC BiCMOS. In: IEEE International Conference on Semiconductor Electronics, 2006. ICSE'06. 2006, pp. A7–12.
60. Tazzoli A, Danesin F, Zanoni E, Meneghesso G. ESD robustness of AlGaN/GaN HEMT devices. In: 29th Electrical Overstress/Electrostatic Discharge Symposium, 2007. EOS/ESD, 2007, pp. 4A.4-1–9.
61. Ma Y, Li GP. A novel on-chip ESD protection circuit for GaAs HBT RF power amplifiers. In: 2002 Electrical Overstress/Electrostatic Discharge Symposium, 2002. EOS/ESD '02. 2002, pp. 83–91.
62. Muthukrishnan S, Iversen C, Peachey N. A novel on-chip protection circuit for RFICs implemented in D-mode pHEMT technology, in 29th Electrical Overstress/Electrostatic Discharge Symposium (EOS/ESD), 2007. EOS/ESD. 2007, pp. 4A.3-1-4A.3–7.
63. Qiang C, Parthasarathy S, Salcedo JA, Liou JJ, Hajjar JJ, Yuanzhong Z. Snapback and post-snapback saturation of pseudomorphic high-electron mobility transistor subject to transient overstress. IEEE Electron Device Lett. 2010;31:425–7.
64. Ersland P, Somisetty S. ESD protection capabilities of GaAs Schottky diodes. In: ROCS Workshop, 2007. Reliability of Compound Semiconductors Digest. 2007, pp. 141–52.
65. Lin CK, Liu J, Sun J, Wang YC. On-chip ESD protection circuit using enhancement-mode HEMT/MESFET technology. Google Patents.
66. Maoyou Sun YL, Yongjin H. Low capacitance ESD protection circuits for GaAs RF ICs. J Electrost. 2006;65:189–99.
67. Shen SC, Caruth DC, Huang JJ. Device technologies for RF front-end circuits in next-generation wireless communications. Proc IEEE. 2004;92:354–75.
68. Barth JE, Verhaege K, Henry LG, Richner J. TLP calibration, correlation, standards, and new techniques. IEEE Trans Electron Packag Manuf. 2001;24:99–108.
69. Cuin Q, Zhang S, Zhao Y, Hou B, Liou JJ. A novel electrostatic discharge (ESD) protection circuit in D-Mode pHEMT technology. In: Compound Semiconductor Integrated Circuit Symposium (CSICS), 2012 IEEE, pp. 1–4.
70. Qiang C, Chia-Shih C, Liou JJ, Hsien-Chin C. Development of a new pHEMT-based electrostatic discharge protection structure. IEEE Trans Electron Devices. 2011;58:2974–80.
71. Muthukrishnan S, Hale C, Peachey N. A novel clamp based ESD protection structure for high power RF Ports in GaAs pHEMT process. In: Compound Semiconductor Integrated Circuit Symposium (CSICS), 2011 IEEE, pp. 1–4.

MIX
Papier aus verantwortungsvollen Quellen
Paper from responsible sources
FSC® C105338

If you have any concerns about our products,
you can contact us on
ProductSafety@springernature.com

In case Publisher is established outside the EU,
the EU authorized representative is:
Springer Nature Customer Service Center GmbH
Europaplatz 3, 69115 Heidelberg, Germany

Printed by Libri Plureos GmbH
in Hamburg, Germany